MW00649920

EXTENDED REALITY IN PRACTICE

EXTENDED REALITY IN PRACTICE

100+ AMAZING WAYS VIRTUAL, AUGMENTED AND MIXED REALITY ARE CHANGING BUSINESS AND SOCIETY

BERNARD MARR

WILEY

This edition first published 2021.

© 2021 by Bernard Marr

Registered office

John Wiley & Sons Ltd, The Atrium, Southern Gate, Chichester, West Sussex, PO19 8SQ, United Kingdom

For details of our global editorial offices, for customer services and for information about how to apply for permission to reuse the copyright material in this book please see our website at www.wiley.com.

Wiley publishes in a variety of print and electronic formats and by print-on-demand. Some material included with standard print versions of this book may not be included in e-books or in print-on-demand. If this book refers to media such as a CD or DVD that is not included in the version you purchased, you may download this material at http://booksupport.wiley.com. For more information about Wiley products, visit www.wiley.com.

Library of Congress Cataloging-in-Publication Data is Available:

ISBN 9781119695172 (hardback)
ISBN 9781119699385 (ebk)
ISBN 9781119699378 (epub)

Cover Design: Wiley
Cover Image: © Anna_leni/Shutterstock

Set in 11/14 pt Minion Pro by SPi Global, Chennai, India

C9781119695172_140421

Printed and bound by CCPI Group (UK) Ltd, Croydon, CR0 4YY

To my wife Claire, and my children, Sophia, James, and Oliver; and everyone who will use the amazing XR technologies to make our world a better place.

CONTENTS

INTRODUCTION

Why XR Is a Trend on the Rise

I've always wondered whether other people see the world in the same way as me. And I mean that literally, not figuratively. Do other people see the color green in the same way as I see it, for example? Do I see things exactly the same as everyone else, or am I experiencing something unique to me? After all, what is reality, anyway? Isn't "reality" different for all of us?

I may never know for sure whether I see the color green in exactly the same way as others. But what I can do – what we can all increasingly do – is embrace this notion of a reality that's unique to me. This is possible thanks to *extended reality* (or XR for short).

XR blurs the boundaries between the real world and the digital world, meaning it can be used to create more personalized, unique experiences. For now, this is mostly used to create immersive experiences in marketing, education, tourism and the like. But in the future, it could extend to all aspects of life as we know it – to the point where each one of us could potentially transform the real world around us into something personalized, using special glasses, headsets, or maybe even contact lenses and implants. Let's say you hate the garish paint job your neighbors have done on the exterior of their home. In the future, your glasses could change it for you, and you'll see whatever color house you choose. Or let's say you see an impressive building and want to know who designed it and when it was built. Your glasses will be able to tell you, overlaying the info directly in front of your eyes (or you'll be able to point your phone camera at the building and see the relevant info onscreen).

Increasingly, our experience of the world will take place in this blurred area between the real world and the digital one. If you think of the time people spend on social media, crafting their online persona, it's clear the line between the digital world and the real one has already become pretty porous. XR will accelerate this. If that sounds a little ominous, it's not. I believe XR is going to change our world and transform our businesses *for the better*. As the examples in this book show, it's already happening.

To be clear, this isn't a tech book. It's not about how to build XR experiences. It's about real-world applications, and the incredible possibilities of XR, now and in the future. It looks at how XR is already being used in practice, across a range of different industries, and what these state-of-the-art applications might mean for the future. I've therefore written the book with business leaders in mind, but hope that anyone interested in this huge tech trend will find inspiring food for thought in these pages.

Why This Book, and Why Now?

As a futurist, it's my job to look ahead, identify transformative tech trends and tell people about those trends as they begin to burst into the mainstream. It's something I've done before with key trends like artificial intelligence (AI) and big data. Given that XR is predicted to become a $209 billion market by 2022,[i] I'd earmarked it as another burgeoning trend to watch closely.

That's to say I planned this book before the coronavirus crisis hit, and started writing while under lockdown in the UK. During lockdown, it became even more obvious that XR is a tech trend rapidly on the rise – and that the technology will now be fast-tracked by many companies.

In the pandemic, our lives moved further online

What was already a trend before COVID-19 quickly became a way of life for many, giving businesses a vital way to maintain connections between people, from the comfort (and safety) of their homes. Pretty much overnight, people who had previously gone to work in an office were conducting daily video calls from home (with increasingly impressive virtual backgrounds), and new tools surfaced that simulate the experience of working in an office environment. Argodesign's artificial window concept is just one example. It's an LCD screen that goes on the wall and looks like a window with the shade pulled down – but if you pull up the shade, you see a colleague (or colleagues) through the "window." You can even chit-chat and make awkward eye contact, just like in a real office.

Virtual conferences are another good example. As traveling to in-person conferences was suddenly no longer an option, virtual conference experiences – like those provided by VirBELA – stepped in to bridge the gap with immersive online conferences, right down to the breakout sessions.

Work, as we know it, may never be the same

Many experts, myself included, believe coronavirus will change the very nature of work, tipping the balance in favor of more remote working. Which means our lives will become ever more digital, and those digital experiences will need to become even more realistic. Interactions between the real world and the digital world will become all the more seamless. The boundaries between the real and the virtual will further blur.

In the future, then, we'll be able to have our business meetings and team-building sessions in whatever virtual settings we want – around a campfire in the middle of a gorgeous wildlife resort, in a futuristic office, on a beach, or even on the Moon. Why not? XR makes anything possible. And we won't even need to leave our homes to do it. You could prepare for that big presentation in front of a virtual audience before you present it in the real world. And after that big presentation, the team could let off steam by going to a (virtual) Rolling Stones gig, or watching a Manchester United or Dallas Cowboys game from your (virtual) corporate VIP box.

Evolving relationships with customers

The pandemic also gave us a taste of how XR will alter the customer experience. Unable to connect with customers in the real world, lockdown presented many businesses with a stark choice: adapt or die. Again, XR provided a way to maintain those connections with customers and give them a unique, memorable experience. One great example comes from Barcelona-based bridal company, Pronovias Group, which launched a virtual showroom and virtual appointments, allowing customers to shop the latest bridal collections at home. Going forward, XR could deliver many more opportunities to immerse customers in the brands they love, and support the in-person customer experience.

A "perfect storm" of technology

There's another reason this book is so timely: we're entering a new industrial revolution – the fourth industrial revolution – where innovation is being driven, in particular, by AI and big data. These technologies feed into and enhance XR technologies, as do other tech trends like 5G, cloud computing and edge computing (processing data close to the source of where it is generated). This perfect storm of technology will aid the development of new XR solutions and make XR experiences even more powerful in the very near future.

Introducing the Extended Reality Spectrum

I delve more into the technology itself in Part 1, but, for now, let's take a brief look at what XR means. XR is in fact an umbrella term for a range of immersive technologies, spanning the ones we already have today – virtual reality, augmented reality and mixed reality – plus those that are yet to be created. In terms of the current technology, we have:

- **Virtual reality (VR)**, which offers an experience that is fully immersive. Here, the user effectively blanks out the real world and enters a computer-simulated environment – typically using a special headset or glasses, like the Oculus Rift headset.

- **Augmented reality (AR)**, which blends the real world and the digital world by overlaying digital objects or information onto the real world. (Think of the addictive Pokémon GO game, where players could "see" Pokémon characters on the street, and you get the idea.) So, while VR creates a simulated environment, AR is very much rooted in the real world. And unlike VR, AR doesn't require specialist equipment – a simple smartphone with a camera will do.

- **Mixed reality (MR)**, which sits somewhere in between the two to create a hybrid reality, where digital and real-life objects can interact with each other. So, for example, the user can move or manipulate virtual elements as if they were right in front of them. This differs from AR, where the user can't interact with the objects or information being overlayed.

Clearly, then, XR represents a spectrum, with some of the technologies being way more advanced and impressive than others. Some require specific hardware, while others harness the capabilities of the average smartphone. The interfaces are constantly evolving, and it's likely we'll experience XR in completely new ways in future. But, across the spectrum, all the different XR technologies have one thing in common: they enhance or *extend* the reality we experience, whether it's by blending the virtual and real worlds together or by creating a fully immersive digital experience that feels as authentic as the real world.

This ability to create more immersive digital experiences or enhance the experience of the real world around us is going to transform many businesses and industries. It will provide companies with new ways to connect and engage with their customers, and improve the customer journey. It will also bring exciting new opportunities to improve business processes, including training, education and hiring.

In short, XR will turn *information* into *experiences*. And this has the potential to change, well, pretty much everything.

The Incredible (And Very Real) Possibilities of XR

In Part 2 of this book, we'll explore real-world use cases from the here and now – compelling examples of how the world's biggest brands are starting to use XR in practice. For example:

- **In everyday life.** Could VR make us better people? That's certainly the idea behind the concept of *virtual embodiment*, which gives users a chance to explore the world from another's point of view. In one example, Courtney Cogburn, an assistant professor at Columbia University, created a VR film called *1000 Cut Journey* that allows viewers to experience the impact of racism on African Americans. Head to Chapter 4 for more inspiring examples like this.

- **In retail.** My teenage daughter is a glasses-wearer. As anyone who wears glasses will know, the traditional way to find frames that suit is to spend ages in-store, trying on many, many different pairs – an unpleasant prospect at any time, let alone during a pandemic! Eyewear retailers like Warby Parker are now using AR to help customers "try on" glasses virtually, using the face-scanning capabilities on their phone. Customers like my daughter can then see a 3D preview of their face "wearing" different glasses, without having to go to a store. Turn to Chapter 5 for more retail and customer engagement examples like this.

- **In training and education.** Both AR and VR are making huge waves in training and education. In one example, Hilton's VR training program uses a combination of computer graphics and 360-degree video to simulate room service, housekeeping and front desk tasks. The idea is to teach corporate team members what it's really like to be a frontline Hilton employee, so they don't inadvertently set policies that make colleagues' jobs more difficult. In Chapter 6, you'll find more use cases from the world of training and education.

- **In healthcare.** Ever had to give a blood sample and endured the squeamish experience of the nurse or doctor struggling to find a vein? The AccuVein AR scanner could change all that. It projects where veins are in order to help healthcare professionals find the patient's veins more easily. Head to Chapter 7 for more healthcare examples.

- **In entertainment and sport.** The potential to create immersive experiences for fans clearly offers huge benefits in sports and entertainment. For example, Oz Sports has launched OZ ARENA, an AR experience that brings fans watching at home *into* live sporting matches, allowing them to personalize their experience and even "appear" at the game in their preferred seat in the stadium. You'll find more entertainment- and sport-related use cases in Chapter 8.

- **In real estate and construction.** With VR, you can tour a property without having to get out of bed. Luxury realtor Sotheby's already provides virtual walkthroughs for iPhone and Android users, and VR is set to transform all manner of real estate sales. With VR, you can even do virtual tours of properties that haven't been built yet. Chapter 9 explores real estate and construction in more detail.

- **In tourism and hospitality.** If you can tour real estate, it makes sense you can tour a hotel or resort to see if it's right for you before you book. Many luxury hotels are already providing virtual tours, such as Atlantis, The Palm in Dubai. Discover more cases from the world of tourism and hospitality in Chapter 10.

- **In industry and manufacturing.** Wiring a jet airliner isn't an easy task. But Boeing is making it easier with AR. Using Google Glass AR technology, Boeing technicians see instructions and how-to-videos in their field of view, and get helpful voice commands – all of which has helped to make the wiring process quicker and more accurate. Head to Chapter 11 for more manufacturing-related examples.

- **In law enforcement and military.** The U.S. Army is using AR technology to help soldiers improve their situational awareness. The technology, which is called Tactical Augmented Reality, is basically an eyepiece that helps soldiers better understand their position and those of others around them, including fellow soldiers and enemies. Chapter 12 explores other military and law enforcement uses.

Having explored the current state-of-the-art in XR, in Part 3 of this book I'll take a look ahead and see where XR might be heading in the future.

Key Takeaways

In this chapter, we've learned:

- XR blurs the boundaries between the real world and the digital world. Increasingly, our experience of the world will take place in this blurred area between the real world and the digital one.

- What was already a burgeoning tech trend suddenly became more important and urgent during the COVID-19 pandemic – a development that will lead many companies to fast-track their XR applications.

- XR is an umbrella term for a range of immersive technologies, including virtual reality, augmented reality and mixed reality – plus those technologies that are yet to be created.

- XR is already being used across a range of industries, including retail, healthcare, manufacturing and many more. This book is packed with real-world examples of XR in action.

I hope this introduction to XR has whet your appetite, and you're now keen to learn more about XR's capabilities. In the next chapter, we'll delve into XR technology in more detail.

Endnote

i. What Is Extended Reality (XR)?; Visual Capitalist; https://www.visualcapitalist.com/extended-reality-xr/

1
WHAT IS EXTENDED REALITY?

Without getting too bogged down in technical details – after all, this isn't a tech book – it's worth spending some time exploring the different technologies that sit under the XR umbrella. Therefore, this chapter gives you a basic grounding in the XR spectrum, including how the various XR technologies work, and what they can do.

A Word About XR Definitions

My goal in this book is to showcase the world of XR, and how XR technologies are changing our lives and our businesses. What I'm not trying to do is rigidly define each type of XR and draw distinct boundaries between the different technologies.

Remember, XR is a spectrum

This is important because XR is still very much a developing field, and it's not always clear where one XR technology ends and another begins. For example, experts can get far too caught up in whether something should be classified as augmented reality (AR) or mixed reality (MR). To me, that just isn't useful, nor is it particularly relevant. At least, not from a business perspective. I imagine you, the reader, want to grasp the potential of XR and understand how it can improve certain elements of your business – and you don't much care where the boundary between AR and MR lies. I make the assumption that you're interested in uses, results and outcomes, as opposed to academic debate.

It's also worth noting that, just as the boundary between the real world and the digital world is becoming more blurred, so too are the boundaries between the different XR technologies. As XR advances, the various technologies that sit under the XR umbrella will become more and more linked, and users will be able to seamlessly move from one technology to another.

So, in the future you may use AR to bring information to life in the real world, then switch to VR to deepen that experience. Say, for example, you're taking a (real-life) holiday on a Greek island. Using AR, you could point your phone at some impressive marble columns and the information onscreen will tell you those columns once formed the entrance to a site where mysterious ancient rituals were performed. Flip on some VR goggles and you could then immediately step into this world and move among the people of Ancient Greece – no toga required! In the final chapter of this book, I talk more about the future of XR, but one of the key developments I expect to see is a more seamless blending of XR technologies.

XR technology is constantly evolving

What's more, this technology will evolve in ways we can't yet imagine. Remember the fairly brief but intense craze for all things 3D a few years back? 3D movies like *Avatar* and *Gravity* blurred the boundaries between the normal moviegoing experience and something altogether more immersive. Then people started buying 3D TVs for their own home, expecting the home viewing experience to move in a similarly immersive direction. But the concept didn't really take off as expected, and manufacturers quietly shelved their production of 3D TVs. Now, holographic displays are beginning to emerge that revive this notion of immersive home viewing and take it in a new direction. Holographic displays are being developed that can project 3D holograms from the screen, without the viewer having to don clunky glasses (a major downside of the previous 3D wave). This shows us how technology is constantly moving forward, toward a future in which everything in our lives becomes more immersive, more digital – but the specifics of how that technology works, what it's capable of, and even what it's called will change. The same sort of thing may happen within the XR spectrum; for example, it's possible that digital displays will be able to project virtual content onto the real world, without us needing special headsets or apps.

All this means precise definitions will likely become less useful as XR evolves and the boundaries between different technologies become more blurred. That's why we shouldn't get too bogged down in definitions of and differences between concepts like AR, VR and MR. What matters is how we can apply the technology in the real world.

That said, in the interest of breaking up the rest of this chapter into manageable chunks, I'll now attempt to create some loose distinctions between AR, VR and MR. Let's start with AR.

Augmented Reality: The Most Accessible of the XR Technologies

For me, AR has the biggest potential in the short term, because it doesn't have to involve a special kit like goggles or headsets. In many cases, a simple smartphone, laptop or tablet, something with a camera and digitally enabled screen, will do. (Saying that, there are specially designed AR glasses, like Google glass, which will crop up in examples throughout this book.)

What is AR?

Whether it's using specially designed glasses or a simple smartphone, AR involves the projection of digital elements – such as information, graphics, animation or images – onto the real world, so that the digital content being superimposed looks like it is part of the physical world. I've already mentioned Pokémon GO as one example of this technology in action; those Snapchat filters that overlay cute animal ears over your own are another basic example. There's also Google's SkyMap app, which tells you about the constellations as you point your smartphone camera at the sky. Or how about the IKEA Place app, which lets you digitally place IKEA's furniture in your room, so you can check out whether it fits (and how it looks in that space) before you buy.

Because the digital element is superimposed onto reality, the user is still very much in touch with the real world in front of them (unlike, say, a VR experience, where the world created around the user is entirely digital). Yet, thanks to the AR projection, the real world has become enhanced – more informative, more entertaining, or more interactive, for instance.

Head-up displays, which project information onto a windshield, are another interesting example of AR in action. The technology was initially developed for fighter jets, so that the pilot could keep looking ahead while accessing relevant info. Now, cars and trucks are beginning to use head-up displays as a safety feature, in order to help reduce driver distraction. These displays project real-time information such as GPS maps or vehicle information either directly onto the windshield itself (in cases where the technology is included in the vehicle as standard) or onto a film that's been added to the windshield (in cases where the technology has been retrofitted). Just as in those fighter jets, the idea is to keep the driver's eyes front and center, giving them the info they need at a glance, without hindering their view of the road ahead.

How does AR work?

AR needs a live camera feed in order to add digital content on top of the real-world elements. The camera feed is what allows the AR system to understand the physical world, so that it can add the right digital content in the right place (a puppy nose over your real nose, for instance). This is all possible thanks to computer vision, also known as machine vision – essentially, a subset of artificial intelligence (AI) that helps machines "see" the world around them and respond accordingly.

Once it has the live, real-time camera feed (be it of a building, the street, your friend's face, or whatever), the AR system then renders digital content on top of the relevant real-life content, making sure it overlaps correctly and is located in the right place. This is updated in real time as the camera feed changes – say, as you're walking down the street holding up your phone.

Stepping into a More Immersive Environment with Virtual Reality

VR offers a far more immersive experience than AR, but, in order to do that, it requires more technology and infrastructure (at the very least, a VR headset). The good news is that this kit is getting lighter, better and less cumbersome. We no longer need heavy headsets with lots of cables that connect to a computer. Now, we can have a light-weight, standalone headset or head-mounted display that doesn't need to be plugged into a main computer. The technology is getting cheaper, too – for just a few dollars, you can get a basic Google Cardboard VR viewer that, along with an accompanying app, transforms your smartphone into a VR device. Of course, for the best VR experience, you currently still need fairly elaborate gear, such as headsets, controllers and speakers. But there's no doubt that the technology is generally shrinking, and getting cheaper and simpler – all of which helps to make VR rapidly more accessible.

What is VR?

While AR is rooted in the real world, VR creates a 3D, 360-degree experience of an artificial, computer-simulated ecosystem. Strap on a VR headset and you're completely transported into this artificial world – whether it's being underwater and exploring a coral reef, walking on the Moon, visiting Ancient Egypt, or whatever. Meanwhile, the real world around you is totally blocked out. Such VR headsets include the Oculus Rift, HTC Vive, GearVR and the previously mentioned Google Cardboard (which is, you guessed it, made of cardboard). These vary in sophistication in terms of how slick and seamless the experience is.

The world of gaming was an early adopter of VR technology, and is perhaps still the first thing people think of when it comes to VR experiences. But, as you'll see in this book, many other industries are now beginning to harness the possibilities of creating fully immersive experiences for customers and colleagues alike.

One recent VR example is the Spatial app. This is a virtual meeting space that lets you meet up with colleagues or friends, whether or not you have a VR headset. If you don't have a headset, you can simply join using a web browser on your phone, tablet or computer. This is an important leap forward because it means people without a special VR kit can still join in the experience. Spatial is also free and open to everyone (a paid-for, enterprise version with enhanced features is also available).

With Spatial, you can meet with others in a beautiful virtual meeting space, and, thanks to virtual avatars – you can take a picture of your face to create your own personalized digital avatar – it feels like you're really in the room together. What's more, your avatar can move around the room and gesticulate as you talk. As you can probably imagine, this is a far cry from the average Zoom or Skype experience, where you're just looking at a wall of 2D faces. Spatial says it has experienced a huge surge in demand – approximately a 1,000 percent increase – in the wake of COVID-19.[i] I'm not surprised. Tools like this will revolutionize remote working.

(As an aside, the use of personalized avatars is particularly interesting to me, and something that we're likely to see a lot more of across various XR technologies. In the future, we could all have different avatars for different digital settings. For example, you could have a smartly dressed avatar for your virtual work meetings. You could have a completely different avatar [animal, human, whatever] for gaming and hanging out with friends online. And you could also have a very realistic avatar, one that accurately reflects your real-life size and shape, which you could use to virtually try on clothes before you buy.)

How does VR work? The super-quick version

Vision is key to creating an immersive 3D environment, which is why special VR headsets are needed. Therefore, a VR headset is, in essence, a small screen (or it could be two screens, one for each eye). Sound effects are also key to creating a consistent, engaging experience, which is where speakers and headphones come into play. Then you have head- and eye-tracking technology to track the user's movements. This may use laser points and infrared LED lights within a headset, or sensors within a mobile phone – or, in very sophisticated systems, special cameras and sensors can be installed in the room to monitor movement.

Merging the Real and Digital Worlds with Mixed (Hybrid) Reality

I've already mentioned how the line between reality and the digital world is becoming increasingly blurred. MR – sometimes referred to as hybrid reality – plays on this notion and takes it to a new level by combining elements from VR and AR. MR is by far the least mature of the three XR technologies featured in this book. However, as we'll discover, companies are already beginning to use MR to solve their business challenges, support new initiatives and improve business processes.

What is MR?

There are lots of confusing definitions surrounding MR and, in particular, some debate over what constitutes MR versus AR. For me, the distinction is this: MR blends components of the digital world with the real world in real time, to the extent that you can interact with the digital elements as if they were real objects. This creates a more immersive experience than straightforward AR. For example, instead of seeing a projection of a digital object on top of the real world (as you would in AR), MR would let you move that digital object with your hands, turn it around to inspect it from different angles, make it bigger or smaller, and so on. With MR, you don't fully block out the real world, as you would in a VR experience. Rather, you're able to experience a virtual environment and the real world at the same time.

One example of MR in action comes from British company BAE Systems, which uses MR to enhance its production of electric bus batteries. Using Microsoft's HoloLens MR headset, BAE workers can project 3D images and instructions onto their workspace, and follow the digital instructions to construct the complex batteries. According to BAE, the use of MR has reduced the time it takes to build batteries by up to 40 percent.[ii]

How does MR work?

MR requires a dedicated MR headset and a lot more processing power than VR or AR. It may also require the use of controllers and motion tracking technology, such as gloves that track your hand movements so you can interact with digital objects.

At the time of writing, the Microsoft HoloLens is the main MR headset on the market, and it comprises holographic lenses, a depth camera, a variety of sensors, plus speakers. With the HoloLens, you look through the headset and see your normal surroundings. But you'll also see holograms (for example, virtual beings, information or objects) overlayed on top of the real world – and, using hand controllers or specific gestures, you can play around with these holograms as if they were real. For instance, you might see a digital to-do list beamed onto your office wall and be able to wipe items off the list as you complete them.

Where Is XR Technology Heading?

As I've already mentioned, in the future I believe AR, VR and MR will all merge together to create more immersive user experiences, where you can move from one device to the next to deepen the experience. Where you can move from an experience that's more rooted in the real world to one that's fully digital. This blending of technology will eventually allow us to see the world however we fancy – to turn the real world around us into whatever we want. Pink trees instead of green. A cartoon avatar instead of your boss. A rainforest instead of a bland conference room . . .

And the technology itself will change. Right now, to get a fully immersive VR experience, you need special gloves or even full body suits to track your movements and simulate the feeling of touch. In the future, everyday cameras will be able to integrate with XR experiences and track our movements. Beyond that, brain–computer interfaces could be used to simulate the feeling of touch, without needing any external technology at all. Then we'll have the integration of smell, and freer movement (thanks to things like omnidirectional treadmills, that let you carry on walking in whatever direction you want).

You can read more about this futuristic vision of an XR-driven world in Chapter 13. For now, the key message is this: although it's obviously helpful to understand what XR technology can do right now, it's vital we remember XR will evolve in ways we can't yet imagine.

Key Takeaways

In this chapter, we've learned:

- XR is a spectrum and, as such, it's not always clear where one XR technology ends and another one begins. This book therefore focuses on real-world applications of the various XR technologies, rather than prescriptive, academic definitions that have little relevance in the real world.

- As XR advances, I believe the various technologies that sit under the XR umbrella will become more and more linked, and users will be able to seamlessly move from one experience to another – for example, moving from AR or MR to VR, and back again.

- AR, which involves the projection of digital elements (such as text or images) onto the real world, has the biggest potential in the short term, because it doesn't necessarily involve special equipment. In many cases, a smartphone is all you need.

- While AR is rooted in the real world, VR creates a much more immersive, completely simulated ecosystem. Strap on a VR headset and you're transported into a 3D, 360-degree artificial environment – while the real world around you is blocked out.

- MR blends VR and AR to create a hybrid reality, in which users can interact with digital elements being superimposed over the real world, as if they were real objects.

I've already set the scene for where I believe XR technology is headed. But what about where it has come from? How did we get to this point, where the line between the real world and the digital one has become so blurry? Turn to the next chapter to trace the evolution of XR.

Endnotes

i. You Can Now Attend VR Meetings – No Headset Required; Wired; https://www.wired.com/story/spatial-vr-ar-collaborative-spaces/
ii. MR Is Leaving AR in the Dust; Iflexion; https://www.iflexion.com/blog/mixed-reality-examples

2
THE AMAZING EVOLUTION OF XR: A BRIEF HISTORY

Having got a basic understanding of how current XR technologies work, now is a good time to pause, and take a look at how we got here.

Tracing the XR Timeline

Let's breeze through the key milestones in the evolution of XR.

The 1800s: A critical discovery

Surprisingly, we can trace the origins of XR as far back as 1838, when scientist Sir Charles Wheatstone first outlined the concept of "stereopsis" or "binocular vision" – which describes the process when the brain combines two different images (one from each eye) to make a single 3D image. This led to the development of the first stereoscopes, devices which used a pair of images to create one single 3D image with the illusion of depth. In today's virtual reality (VR) systems, stereoscopic displays can be used to enhance the feeling of immersion by bringing a sense of depth to digital images.

The early 1900s: Forecasting the future of VR

The first prediction of VR as we know it came in 1935 when American science fiction writer Stanley Weinbaum published *Pygmalion's Spectacles*, in which the main character is able to explore a fictional world thanks to a pair of goggles. In the story, the wearer experiences not just virtual sounds and sights, but also taste, smell and touch. Looking at today's advanced VR systems, and where VR might be going in future, Weinbaum's prediction is startlingly accurate.

The 1950s, 1960s and 1970s: The first VR and AR experiences

The first VR machine, named Sensorama, was created in 1956 by cinematographer Morton Heilig. It was essentially a movie booth that combined 3D, color video (using a stereoscopic screen), audio (from stereo speakers), smells (from scent producers) and vibrations (from a vibrating chair). Six short films were developed for the booth, with the goal of fully immersing the viewer in the movie.

In 1960, Heilig patented the first head-mounted display, called the Telesphere Mask, which combined stereoscopic 3D images with stereo sound. Heilig's headset didn't include motion tracking, but the following year, engineers from Philco Corporation created Headsight, the first motion-tracking headset. Headsight was designed to help the military remotely assess hazardous situations, with the user's head motions being mimicked by the remote camera.

Also, in the 1960s, computer scientist Ivan Sutherland presented a paper outlining his concept of the "Ultimate Display," a virtual world so realistic that the user wouldn't be able to differentiate it from reality. This is widely considered the blueprint for modern VR.

Not forgetting augmented reality, the first AR headset was created by Harvard professor Ivan Sutherland in 1968. Called "The Sword of Damocles," the headset displayed computer-generated graphics that enhanced the user's perception of the world.

In the 1970s, MIT created Aspen Movie Map, a computer-generated tour of Aspen that allowed people to wander, virtually, through the streets of Aspen. Created from photographs taken by a car driving through the city – incredibly, predating Google Street View by decades – this program was perhaps the first to show how VR could transport users to another place and perfectly replicate that place's streets and buildings.

The 1980s, 1990s and 2000s: Gaming adopts VR

New technology emerged in the 1980s to support the VR experience, including the Sayre gloves, wired gloves that monitored the user's hand movements – the origin of gesture recognition. Plus, in 1985, VPL Research Inc., the first company to sell VR goggles and gloves, was founded. Jaron Lanier, one of the founders of VPL Research, coined the term "virtual reality" in 1987. The term "augmented reality" was coined in 1990 by Boeing researcher Tom Caudell.

The early 1990s saw the introduction of VR arcade machines, like the SEGA VR-1 motion simulator. Then, in 1995, Nintendo launched its Virtual Boy 3D video games console, the first portable console with 3D graphics. But being on the clunky side, and with monochrome graphics only (so no color), it wasn't a big success and ended up being discontinued a year later. Crucially, in the mid-1990s, we saw the release of affordable VR headsets for home use – such as the Virtual IO I-Glasses, which came with head-tracking technology.

In terms of AR, the 1990s brought game-changing technology to sports broadcasting. In 1998, Sportsvision broadcast the first live NFL game with the yellow yard marker (a yellow line overlayed on top of the live camera feed). This idea of overlaying graphics over the real-world view quickly spread to other sports and indeed other areas of TV broadcasting.

The early 2000s was a quiet time for XR technologies, but the introduction of Google Street View in 2007 extended the idea first posed by the Aspen Movie Map, making it possible for viewers to experience a different city.

2010 to 2020: XR technologies gain momentum

The next big leap came in 2010 when Palmer Luckey created the prototype for the Oculus Rift VR headset – at the tender age of 18, no less. The headset's 90-degree field of vision and use of computer processing power was totally revolutionary, and served to reignite interest in VR. In 2012, a Kickstarter campaign for the Oculus Rift headset raised $2.4 million. Oculus VR, the company founded by Luckey, was acquired by Facebook in 2014 for around $2 billion, and this is when VR really started to pick up momentum. That same year, Sony and Samsung announced they were getting in on the VR headset act, and Google released its first Cardboard device, a low-cost cardboard VR viewer for smartphones.

Also in 2014, Google unveiled its Google Glass AR glasses, which overlay digital information onto the real world. Users wearing the glasses – who were quickly and cruelly dubbed "glassholes" – could communicate with the internet via their glasses, and access applications like Gmail and Google Maps. (As an aside, mainstream adoption of Google Glass didn't exactly go the way Google expected, and consumer response was lukewarm at best. But, undeterred, Google pivoted to enterprise editions of Google Glass, encouraging employers to kit out their workforce with AR glasses. As we'll see in Part 2, this strategy appears to be paying off, and a variety of employers are now using AR in the workplace to improve productivity and accuracy.)

In 2016, Microsoft released its HoloLens headset, which took the idea of AR to a new level by creating a more interactive experience (hence, the phrase "mixed reality," or MR). It was also the year in which people all over the world became addicted to the AR-driven Pokémon GO game – which, seemingly overnight, bought AR into the mainstream in a way that Google Glass never had.

By the end of 2016, hundreds of companies were developing VR and AR experiences – and we're not just talking about tech and gaming companies. The BBC, for example, created an immersive, 360-degree video of a Syrian migrant camp, and the Washington Post created a VR experience of the Oval Office.

In 2017, we saw an early application of AR in mainstream retail, with the release of the IKEA Place app, which lets users see how furniture would look in their home before they buy. By 2020, the use of VR, AR and MR had quickly extended to a wide range of industries, beyond retail (see Part 2 for case study examples from across different industries). This widening of XR applications is critical, proving that XR is no longer seen as just a gaming/entertainment thing.

This rush to develop XR experiences across many different industries is further driving interest and investment in XR equipment, like VR headsets. At the time of writing (2020), many more companies have developed or are developing their own VR and AR hardware, including Apple, Google, Huawei and HTC. Which brings me to . . .

Rapid Evolutions in XR Hardware

XR hardware is now developing extremely quickly, across various different displays. Let's take a look at the main advances.

VR equipment

One of the biggest limiting factors with VR has been the need for clunky headsets connected to a computer. Nowadays, we have wireless headsets that are small, light, comfortable, and incredibly powerful. Take the self-contained Oculus Quest headset as an example. Or the Feelreal headset, the first patented multisensory VR mask capable of simulating smells, hot and cold breezes, vibrations and even punches! Importantly, VR devices are now becoming more affordable. We have VR headsets/viewers ranging from just a few dollars (Google Cardboard) to mid-range VR headsets priced at a few hundred dollars, like Oculus Quest. Of course, at the top end, things are still pretty pricey – Microsoft's HoloLens 2 MR headset retails, at the time of writing, for a cool $3,500.

Gesture recognition – where the VR system is able to interpret human gestures – is also improving at pace. While you used to need special equipment, usually gloves or controllers, to achieve gesture recognition, now we have vision-based gesture recognition using cameras, plus the development of full-body motion-tracking VR suits that look surprisingly comfortable (like the TESLASUIT). And at the cutting edge of VR experiences we have CAVE environments, cube-like spaces with projectors, tracking sensors and speakers that – combined with a headset – help to create a truly immersive experience.

AR apps and glasses

Meanwhile, in AR, impressive apps are being developed that harness the capabilities of the average smartphone, such as SketchAR, which helps you recreate a digital photograph as a hand-drawn sketch (by projecting the image onto paper, so you can trace the lines). Ink Hunter lets you try out tattoos (both premade designs and your own designs) on your body before you take the plunge. And Mondly improves the experience of learning a new language by digitally placing an AR teacher in the room with you.

It makes sense, then, that phones are now routinely being built with AR and VR uses firmly in mind. For example, the next generation of iPhones are expected to include LiDAR scanners – the same technology that helps self-driving cars scan their surroundings and detect obstacles like pedestrians and cyclists. Such technology will improve the iPhone's ability to create 3D maps, and improve object "occlusion" (where digital objects disappear behind real-world objects). The idea is this will make games and apps all the more realistic and impressive. For Android phones, Google has its ARCore platform, enabling Android phones to sense their environment in order to improve AR experiences. Huawei's AR Engine does a similar thing. Bottom line: our phones and their operating systems are increasingly being designed to support a range of XR experiences.

The ability of smartphones to support AR is great because it means users don't need to buy a special kit, like glasses. That said, AR glasses do have certain benefits over smartphone apps. After all, there are occasions when operating a cellphone just isn't desirable or safe, in which case AR glasses will be more suitable. For example, an AR app that translates a menu from Japanese into English before your eyes is great – holding your phone up to a menu while dining out on holiday is no great hardship. But what if you're an engineer up a ladder fixing something? Holding up a cellphone is hardly ideal in that scenario – but AR glasses can deliver the instructions you need in front of your eyes, while you keep your hands free. Those who are interested in AR glasses could soon be in for a treat; Apple is reportedly developing its own AR glasses to rival Google Glass.

Other types of XR displays

The heads-up displays I mentioned briefly in Chapter 1, where information is projected onto a windscreen, are another example of how XR devices have evolved beyond headsets, glasses and smartphone apps. Initially developed by the military for use in fighter planes, heads-up displays are now becoming increasingly commonplace in our cars. Almost every luxury car brand has at least an optional heads-up display feature, showing things like GPS directions, speed, etc.

Currently, heads-up displays can only project a two-dimensional image or graphic. What will drive heads-up displays into the mainstream will be the incorporation of 3D images that give a sense of depth. This will make the projection of things like lane markers and GPS directions all the more accurate and helpful for drivers because, instead of seeing a basic 2D directional arrow, you'll be able to see a 3D arrow that curves off into the distance to show exactly where you need to turn. This demonstrates how we can expect exciting advances in XR technology, beyond the obvious AR and VR headsets/viewers. In fact, this sort of technology could be incorporated into many everyday objects. Like mirrors.

Several companies are making serious headway in the development of "smart mirrors," and in the near future these could be incorporated into settings such as retail dressing rooms, barbers and beauty salons. Rather like an AR app, but hanging on the wall in front of you, these mirrors can adjust how you look digitally, including the clothes you're wearing, your hairstyle and even your hair color. The Wella Professionals Smart Mirror, which was honored by the CES 2019 Innovation Awards, is one good example, providing a 360-degree view of the client's virtual, augmented hairstyle. Thanks to facial recognition, the mirror can retrieve the customer's past styles, and it can also provide a curated feed of recommended styles according to current trends and classic styles.

An Evolution Fueled By Other Tech Trends

XR has come a long way, but if we look at how other technology trends have developed in recent years, it's clear there's still a lot of exciting stuff to come. After all, the rapid XR evolution seen over the last few years has undoubtedly been enabled by developments in other technology. Artificial intelligence, cloud computing, internet speeds, screen technology, camera resolution, and so on are all accelerating forward, which, in turn, feeds the evolution of XR technologies. As other technologies advance, so too will XR.

AI, for example, will be critical in the move toward multisensory XR experiences that incorporate more intuitive inputs, like gestures, voice and touch. AI will also allow for greater personalization of XR experiences in real time – after all, learning from users and pre-empting what they want is what AI does best. This capability will help make XR experiences all the more powerful. As a simple example, a beauty retail app that lets you see how different lipstick shades look on your face could learn your color preferences and automatically show you the products most suitable.

Then there's the continual increase in computing power. Delivering an immersive VR experience takes a lot of computing grunt. Cloud computing meant we could push that computing load into the cloud, instead of having a headset wired to a powerful computer – a move that has drastically improved the user experience. And now edge computing – where the computing is done closer to the device itself (as opposed to being spread across countless distributed computers in the cloud) – is helping to further improve that experience. Thanks to edge computing, we're already seeing cellphones rapidly become more powerful and more responsive. So too will VR headsets and AR glasses.

And with 5G around the corner, super-fast mobile networks will further boost the potential of XR. We'll have faster network speeds, greater bandwidth and the ability to connect more devices to networks than ever before – ultimately enabling XR devices to become even more portable. In other words, fast, reliable 5G networks should enable us to stream XR experiences whatever our location. This, in turn, will result in even cheaper portable headsets/viewing devices and more realistic simulations.

Key Takeaways

In this chapter, we've learned:

- The early origins of XR can be traced back to the 1800s, yet the most dramatic and rapid advancements have come in the last few years.

- The evolution of XR has seen VR headsets get smaller, lighter and more affordable. Meanwhile, our phones are increasingly being designed to support XR experiences, particularly AR apps. Other types of displays are also being developed, including heads-up displays and smart mirrors.

- Progress in XR has been driven by other key technology trends, including AI and cloud computing. The introduction of 5G networks will further drive the evolution of XR by making XR hardware faster, more responsive and more portable.

It's clear XR technologies have come a long way, but XR isn't without its challenges. In the next chapter, I'll explore some of the pitfalls and disadvantages of XR that businesses need to be aware of.

3
CHALLENGES WITH XR

XR is moving beyond hype and beginning to demonstrate real value. But XR isn't without its problems. Far from it. As we'll see in this chapter, there are many personal and societal risks of XR, particularly with the more immersive XR technologies; hence, some of the issues highlighted in this chapter will be more relevant to virtual reality (VR), which is highly immersive, as opposed to augmented reality (AR) or mixed reality (MR). However, that risk profile may change over time as the line between XR technologies blurs and the whole spectrum of XR technologies becomes more immersive.

Let me stress that the benefits of XR outweigh the potential downsides. But the onus is on organizations, regulators and society as a whole to ensure the balance doesn't tip the other way. Providing we can embed notions like ethics, responsibility and trust in XR technologies, many of the issues I raise in this chapter can (and hopefully will) be overcome in time. After all, we're only just at the beginning of the XR revolution, and the technology is still evolving. That's why now is the time for business leaders to educate themselves on the potential pitfalls, so that they can ensure their organizations get maximum value from XR, while minimizing the risk of harm.

Legal and Moral Concerns

The technology is advancing faster than our legal systems can cope with, leaving regulators playing catch-up. We don't have clear laws on what's acceptable and unacceptable in virtual environments, or even which jurisdictions they come under. (For example, in the future, if I'm physically at home in England, but I'm exploring a virtual environment hosted by a Chinese provider, will I be subject to English laws or Chinese?) Let's look at some of the unanswered legal and moral questions surrounding XR technology.

Can a virtual act be a crime?

In the future, what happens if someone commits what we would normally consider a crime in a virtual environment? What if two people are immersed together in a virtual environment, and one of them assaults the other in that virtual space. Will that be considered a crime? If we base our consideration on video games then the answer is of course not. Many of us indulge in video games where we beat up or shoot our fellow gamers. But the difference with XR technologies is they are much more immersive than the average video game experience. What if, in our hypothetical virtual situation, the participant who is assaulted suffers trauma because the event felt so real. Is it a crime then?

These questions will become more urgent as the technology evolves. Take haptic suits as an example. These suits bring a sense of touch to virtual experiences – so users can feel sensations that are generated in the simulated world. As these evolve, we will be able to realistically feel the touch of other users, potentially opening users up to highly traumatic experiences. It's not unthinkable, for example, that one user would touch another user's digital avatar inappropriately – which that person would then feel in the real world through their haptic suit.

Children are particularly vulnerable when it comes to immersive technologies, and not just in the sense of being vulnerable to exploitation. Research has shown that children can easily confuse what's real and not real – to the extent that children can acquire false memories in a virtual experience.[i] Potentially, this means if a child experiences or sees something traumatic in a virtual world, they could remember it as having taken place in the real world. The same research also shows that virtual avatars and characters are more real and influential than TV characters to children, which is something that might give any parent cause for concern.

Digital theft is another possibility. As we spend more of our lives in the digital world, and accumulate digital possessions and currency in that virtual space (as in the movie *Ready Player One*), it's feasible that other users could steal those virtual items that have become precious to us. (I talk more about XR security issues later in the chapter.) Will the law evolve to cover these sorts of crimes?

These are questions that will need to be answered. Ideally, we would have a global code of conduct for XR technologies, like the codes of conduct that are emerging in the field of artificial intelligence. We need something, ideally universal, that governments and organizations can commit to, to ensure virtual environments are safe spaces for all. But, frankly, we're a very long way away from that happening. For now, the onus is on XR providers to create their own idea of what's acceptable and what's not, and regulate what goes on within their virtual experiences.

Murky moral practices

It's not just a legal question, this conundrum on what should and shouldn't be allowed in a virtual environment. It's a moral issue, too. XR creates the potential to cross a lot of moral boundaries that certainly make me uncomfortable. In theory, it will be possible for someone to render a highly realistic avatar of their neighbor or colleague or friend and then have sex with them. Should that be allowed? It's immoral, sure. But is it *wrong* to commit immoral acts in a virtual world? In my view, if something isn't allowed in the real world (like having sex with someone without their knowledge), it shouldn't be allowed in the virtual world.

Others might argue that it's Orwellian to police people's thoughts and imaginations. But let me ask you this, would you be happy with someone having sex with your digital doppelganger? Your daughter's?

There is already some evidence that pornography plays a role in the sexual violence experienced by some women.[ii] If users engage with pornographic content in a highly immersive, realistic environment, where they are "acting out" the interaction in first-person style, there is a concern that this could further raise the risk of sexual violence in the real world. Then there is the possibility of users carrying out sexual acts that are illegal in the real world – such as rape, sex with minors or sex with animals.

The danger with immersive technologies is they can allow people to act out whatever they want, seemingly without any real-world consequences. Some might say this is fairly harmless. But, increasing the risk of violence is a clear real-world consequence. How will we protect against this risk?

Desensitization is another concern. There is a real danger that spending a lot of time in virtual worlds desensitizes people to real-world issues. Say you spend eight hours in a virtual world where you can do whatever you want. Say you do that regularly. Do you think that might alter your perception of the real world? That's the danger of immersing people in an environment where morality is a gray area, and the normal rules maybe don't apply. The worry is that people who spend a long time in virtual experiences, specifically virtual environments where there are high levels of violence, could end up desensitized to acts of violence. This would mean they're no longer affected emotionally by extreme acts of violent behavior in the real world. As a result, they may fail to show compassion or empathy. How will we ensure the next generation doesn't end up desensitized to the horrors of war, crime, and so on?

In addition, there is the issue of fake news and information seeming more real when it's consumed through an immersive experience. Particularly if a user spends a lot of time in a virtual world, they may find it harder to distinguish between reality and falsehoods. Forget current concerns about fake news – in the future, VR experiences could be used to disseminate highly believable but ultimately fake information, and manipulate users' opinions and behavior.[iii] This makes it all the more important that organizations create and enforce ethical use policies for their digital environments. If they don't, they could see a huge backlash, just as Facebook regularly has to answer to the spread of false information on its platform.

Access for the few, not the many?

Another moral concern is that XR technologies may widen the gap between the haves and the have-nots. The costs of purchasing XR hardware will naturally exclude some people, potentially exacerbating existing social divisions. For instance, XR has a lot of potential to revolutionize education and provide enriching educational experiences for children – but how likely is it those opportunities will be available to all? Not very. Another concern stemming from this is that users who spend a lot of time in virtual environments, where they can have and do whatever they want, will increasingly end up disengaged from social inequalities.

Privacy and Security Concerns

Like most new technologies, XR brings with it some significant challenges related to personal data. Except, in this case, the data can be *extremely* personal. Let's briefly look at the key privacy and security concerns.

Taking the term "personal data" to a new level

These days, we're all used to forking over our personal data in return for products and services that make our lives more convenient. Social media activity, search history, purchase history, viewing habits and credit card details are one thing. But, due to the highly immersive nature of the environment, XR technology has the potential to capture much more sensitive information, including our most intimate behaviors and thoughts.

To enhance the experience, XR technologies are constantly gathering data such as physical location, body movements, reflexes, and potentially even eye movement patterns and voice. This is all necessary to better articulate your "movements" in that digital space and generally make the experience more efficient. But there's no doubt that this is high-risk data.

What happens to the data about these intimate behaviors, and how much control will users have over who has access to the data? Might it be given to advertisers, for instance? Could it be used as evidence of your character in relation to legal proceedings in the real world? To what extent is the information vulnerable to cybertheft and manipulation?

It's in your eyes

Let's take eye-tracking technology as an example. When embedded into VR and AR headsets, eye tracking enables companies to collect highly personal data on your unconscious responses to visual cues (whether virtual, as in the case of VR, or real-world cues, as in the case of AR).[iv] Now, it's important to note that eye tracking brings significant benefits to XR. For one thing, it can help the system focus on what the user is interested in targeting. It also helps to reduce graphics processing (basically only the parts of the image that you're directly looking at get the highest resolution and quality), which means less lag for the user. But we must be clear that the information gathered is open to exploitation.

Patterns in our eye movements show what we're aware of and focusing on at any given time, which gives an insight into our preferences and thoughts, rather like an unconscious "like" button. So if you're walking down the street and your gaze lingers on a particular car, that information could be mined by advertisers to serve up related ads. If that seems harmless, how about this: your eye movement data could give away your sexual orientation and, specifically, who you're attracted to. It could even potentially be used to assess your mental and emotional state.[v] How do you feel about companies knowing all that?

If this all seems a bit too futuristic, consider that eye-tracking technology is already being embedded in VR headsets. The HTC Vive Pro Eye was the first major VR headset to include eye-tracking technology, and the Pico Neo 2 Eye is the first smaller-scale VR headset with eye tracking.[vi] We can almost certainly expect mainstream headsets like the Oculus Quest will be heading in this direction very soon.

Going forward, as we spend more time using these headsets (for example, when we're all walking around with AR glasses on), the potential to understand more and more about users is pretty alarming. Clearly, this takes us to a new level of privacy concerns, which means new thinking is needed to address these issues. My hope is that eye data will eventually be considered as personal as other health data and afforded the same security and privacy protections. At the very least, it shouldn't be stored or shared with third parties without informed user consent.

The risk of identity hacking

In the near future, it will be possible to make a virtual "copy" of yourself that looks and talks just like you (and, thanks to artificial intelligence, can even understand and reflect your personality). Not only could these avatars be used to create new forms of identity-related crime, they could also be vulnerable to ransomware and extortion risks.

We already know how scarily realistic some fake photos and videos are. By hacking someone's digital identity, fake content could become even more powerful and plausible. If someone is (falsely) depicted in an XR environment as doing something criminal or immoral, this could have a serious detrimental impact on their standing in the real world – particularly for people in the public eye.

Our personal eye data could be particularly vulnerable to criminal use. Microsoft's HoloLens2 headset now comes with something called Iris-ID, meaning you can log in by scanning your eyes. Assuming this highly personal biometric data remains protected, it can be more secure than passwords.[vii] But that's a big assumption in this age. Combine iris data with other personal data – your name, credit card information, age, physical appearance and even voice – and users could be left exposed to their entire identity being hijacked. After all, you can reset a password, but you can't change your eyes. One breach of this data and the user's life could get very difficult. How will companies defend against this?

Health Concerns

In addition to the legal, moral, privacy and security issues, there are concerns that use of highly immersive technologies could be bad for our physical and mental health.

Introducing the VR hangover

Users who spend a lot of time in VR sometimes report nausea, dizziness and disorientation. This is the result of the brain receiving mixed sensory signals (such as your eyes registering movement in the digital environment, while your inner ear knows you're standing still in the real world – incidentally, mixed signals like this are why we experience motion sickness on boats and in cars). The brain assumes the contradictory signals are the result of the body being poisoned and triggers feelings of nausea in the hope that you'll throw up the poisonous substance.

Known as a post-VR "hangover," this type of motion sickness can affect different people in different ways. Some experience no discomfort at all while others find even a short experience triggers aftereffects. Experiences with perceived g-forces (like a racetrack with sharp, fast turns) can trigger the worst sickness. And the effects can, in some cases, last up to a week.[viii]

VR can also cause eye problems, such as eyestrain and distorted or blurred vision[ix] – because, unlike in the real world, everything you're seeing is just a few centimeters from your face, which teaches the brain that long-distance focus is no longer needed. And because the technology is still relatively young, we don't yet know what the long-term impacts on users' eyesight will be.

The science on post-VR physical symptoms is still evolving, and no doubt there will be ways for developers to work around these issues. But we need a lot more research before we can fully understand and mitigate against the physical risks.

Post-VR sadness

Some users have also reported problems settling back into the real world – experiencing a disturbing feeling that the real world isn't real, or just isn't as good as the virtual world. In other words, the real world can seem, well, a bit of a let-down after spending time in an environment where you can be whomever you want and do whatever you want. Sometimes referred to as "post-VR sadness," sufferers report feeling detached from reality, sometimes for days or weeks.[x,xi] On top of feeling like the world isn't real (derealization), some users experience depersonalization, or the feeling that oneself isn't real. Rather like having an out-of-body experience.

As VR becomes more popular and immersive, and users spend more time in virtual environments, we seriously need more industry research to determine the effects on people's moods and feelings of anxiety or depression. We also need to fully understand the extent to which virtual acts can inflict real-world trauma and pain on users. In particular, there may be lasting psychological consequences of immersive environments that simulate harrowing or unpleasant experiences.

Cyber-addiction

There is already some evidence that people can get addicted to technology.[xii] It's not unreasonable to assume that the more immersive the technology, the higher the risk of users becoming addicted. The natural consequence of this is, obviously, people spending increasing amounts of time in the virtual world, as opposed to the real one. There is a danger, then, that the virtual world could increasingly replace the real world – especially for users who believe their virtual world, and their persona in the virtual world, is better than reality. Imagine you spend a lot of time in a virtual world that is far more fun than the real one, and where your virtual persona is richer, better looking and more respected than you perceive your real one to be. Isn't there a danger you might begin to prefer that virtual world? The worry is, the more time people spend in the virtual world, the harder it becomes to adjust to reality.

The extreme outcome of this reminds me of the fictional Better Than Life game in the sci-fi comedy series *Red Dwarf*. Using a headset with probes that worm into the brain, Better Than Life players enter a fantasy world that fulfills their every dream. The game is so addictive, choosing to leave is almost impossible – meanwhile, in the real world, players' bodies wither away and they die.

Of course, that's an extreme, far-fetched scenario, but there is a reasonable concern that over-dependence on XR technologies could negatively impact users' mental health and well-being. We could potentially see new mental health disorders emerge as a result.

The Need for Responsible XR

Let me stress again that the benefits of XR outweigh the concerns outlined in this chapter. But I do believe it's vital that business leaders are aware of the potential pitfalls and consider the ethical, responsible applications of XR. It's far better to consider and plan for this at the start of your XR journey than try to "retrofit" responsibility into your systems further down the line. And believe me, just like the backlash over data privacy infringements, irresponsible XR practices will draw public and regulatory ire in the future. Take proactive action now and you'll save yourself a lot of future hassle and reputational damage.

Ideas on responsible, ethical XR are still emerging, but sound principles include:

- Build a culture of responsibility, where everyone in the company is encouraged to weigh and question the ethics of new technologies.

- As much as possible, ensure the technology is as inclusive and affordable as possible – especially if your offering is designed for educational or societal use.

- Always be upfront with users about what data you're gathering, and, where possible, give them the option to opt out. As much as possible, try to anonymize data. Where someone's personal data are absolutely vital, you'll need to take all the same data security measures as you would for any other business-critical process.

Key Takeaways

In this chapter, we've learned:

- There are many personal and societal risks of XR, particularly with the more immersive XR technologies, like VR.

- There are many legal and moral gray areas, including whether a virtual act can be considered a crime, and what constitutes acceptable behavior in a virtual space.

- XR systems have the potential to collect highly personal data, even down to what the user is thinking and feeling. Clearly, this data is open to misuse, potentially even to the extent of identity hacking.

- In addition, we don't yet fully understand the health implications of using XR technologies, including the physical aftereffects (like motion sickness), impacts on eye health, and lasting psychological impacts. Cyber-addiction is another major concern.

- Business leaders must be aware of these issues and work hard to ensure their XR applications are ethical and responsible. Trying to "retrofit" responsibility into your systems further down the line is not a smart strategy.

Now that we've explored the technology itself, and its potential pitfalls, let's see how organizations across a range of industries are delighting customers and improving business processes through XR technology.

Endnotes

i.   Virtually True: Children's Acquisition of False Memories in Virtual Reality; *Media Psychology*, Volume 12, Issue 4; https://www.tandfonline.com/doi/abs/10.1080/15213260903287267?tab=permissions&scroll=top&

ii. Exploring the Connection Between Pornography and Sexual Violence; *Violence and Victims*, Volume 15, Issue 3; https://www.ncjrs.gov/App/Publications/abstract.aspx?ID=187015

iii. The Era of Fake Videos Begins; The Atlantic; https://www.theatlantic.com/magazine/archive/2018/05/realitys-end/556877/

iv. The eyes are the prize. Eye tracking technology is advertising's holy grail; Vice; https://www.vice.com/en_us/article/bj9ygv/the-eyes-are-the-prize-eye-tracking-technology-is-advertisings-holy-grail

v.   Privacy Issues in Virtual Reality: Eye Tracking Technology; Bloomberg; https://news.bloomberglaw.com/us-law-week/privacy-issues-in-virtual-reality-eye-tracking-technology-1

vi. Eye tracking is the next phase of VR, ready or not; Cnet; https://www.cnet.com/news/eye-tracking-is-the-next-phase-for-vr-ready-or-not/

vii. The eyes are the prize. Eye tracking technology is advertising's holy grail; Vice; https://www.vice.com/en_us/article/bj9ygv/the-eyes-are-the-prize-eye-tracking-technology-is-advertisings-holy-grail

viii. Motion sickness and the VR "hangover": What you need to know; Medium; https://medium.com/@ThisIsMeIn360VR/motion-sickness-and-the-vr-hangover-what-you-need-to-know-4d6cb23af121

ix. Developer warns VR headset damaged his eyesight; BBC News; https://www.bbc.com/news/technology-52992675

x.   Virtual Reality Can Leave You with an Existential Hangover; The Atlantic; https://www.theatlantic.com/technology/archive/2016/12/post-vr-sadness/511232/

xi. Effects of virtual reality on presence and dissociative experience; CyberPsychology & Behavior, Volume 9, Issue 6; https://www.researchgate.net/publication/278206370_Effects_of_virtual_reality_on_presence_and_dissociative_experience

xii. Your social media apps are as addictive as slot machines; The Next Web; https://thenextweb.com/contributors/2018/03/25/social-media-apps-addictive-slot-machines-similarly-regulated/

4

XR IN EVERYDAY LIFE AND BUSINESS

From this point onwards, I'll be sharing real-world examples of XR across various different industries and categories, including retail, training and education, and entertainment. Yet, there are some fantastic XR examples that don't neatly fit into such clear-cut categories. Therefore, this initial example chapter is something of a catchall – a collection of interesting and inspiring XR uses from everyday life and business. Consider it a general showcase of what's possible, before we delve into the more specific categories and industries coming up later in the book.

XR in Everyday Life

AR is generally the more accessible XR technology, because you don't need to purchase special hardware to enjoy AR experiences. In fact, a wealth of AR is already available on the average smartphone and tablet . . . .

AR apps are everywhere

AR applications have been around for a while, but they really caught the public's imagination with the launch of Pokémon Go in 2016. Now we have a wealth of AR apps that are designed to help, educate and entertain.

How about an AR app that can solve that impossible Sudoku puzzle for you? Sudoku puzzles are notoriously tricky, sometimes taking hours to complete (that's if you don't give up in frustration before then). The Magic Sudoku app uses machine learning to solve even the most devilishly difficult grid in seconds. Simply aim your camera at the puzzle, and the answers are overlayed on screen for you. It kind of defeats the purpose of doing puzzles in the first place – exercising your brain, passing the time, feeling the satisfaction of completing something – but it shows what AR can do.

For younger folks, Disney researchers have explored using AR to bring kids' coloring books to life. Using a tablet or phone screen, the characters on the page are projected in 3D – a process Disney called "live texturing of augmented reality characters from colored drawings."[i] So, a child could color in the character on the page, inspect their work in 3D using a mobile device, and watch their character stand and wobble around. The texture of the 3D character perfectly matches the texture of the child's coloring lines.

Another great AR use for kids comes from Google, which began adding AR animals to internet searches in 2019. So if you searched for a bear or tiger on an AR-enabled phone or device – that's ARCore-supported Android devices or iOS 11 and up – the animal would "appear" in front of you, overlayed on screen in front of whatever else is there in real life. Apart from being, well, just really cool, this is useful as an educational tool because it helps children understand animals in more detail and gain perspective on their actual size. But it gets better. Now, Google has gone a step further and added a herd of 10 dinosaurs to its AR searches, meaning your child (or you!) can search for a T-Rex on your AR-enabled device and watch it appear in your living room, garden or wherever (scaled to size, of course). My kids absolutely love this.

Speaking of Google, we're all familiar with Google Translate in its most basic form. But Google Translate's Word Lens feature has taken instant translation to a new level. These days, all you have to do is fire up the Google Translate app, point your camera at a sign, menu or whatever it is you want to translate, and the app provides an instant translation, overlaid on live video feed right there on the screen – often using the exact same typeface as the original sign or menu. This works in a range of languages, including Japanese, and you don't need an internet connection for it to work. This live overlaying of translated text is all possible thanks to AR.

Snapchat filters are another AR feature that many of us are familiar with, allowing us to add, shall we say, quirky features like dog ears, glasses and big baby eyes to our faces. There are even Snapchat filters for your dog, too, which enable you to add glasses, antlers, cartoon faces and so on to pictures of your four-legged friend.

I talk more about customer engagement in Chapter 5, but this use of AR filters is becoming an increasingly popular way for brands to entertain and engage with their fans and customers. Thanks to Facebook and Instagram's Spark AR Studio, anyone can create and share their own filters, and a range of brands have leapt on the opportunity. In fact, over one billion people have already used filters powered by Spark AR.[ii] Taco Bell was one of the first brands to catch on to the potential of AR filters, creating a filter that replaced users' heads with a giant taco (complete with Taco Bell's logo, naturally). Coca Cola is another prominent user of AR filters, creating a Christmas-themed filter of a polar bear (holding a bottle of Coke and wearing a branded scarf) that fans could pose with.

Why are brands excited about AR filters? Well, we already know that visual content like images, videos and gifs are more eye-catching than purely text-based content. AR builds on this and makes image and video content even more immersive and engaging, thereby increasing the amount of time customers spend engaging with brands. (According to some studies, AR-based marketing achieves an average engagement time of 75 seconds, compared to just 2.5 seconds for standard

TV ads.)[iii] It makes sense when you think about it – the very nature of AR filters mean you have to spend more time and attention posing with the filter, trying different angles and making funny faces before you capture the perfect shot. Then what do you do? You share it with your friends, of course. . . . No doubt we can expect more and more brands to explore filters as a way to boost customer reach and engagement.

Another example of an interesting AR app comes from WallaMe, which lets users create, hide and share messages in the real world. You take a picture of something around you, say, a building or surface, then add a written message, sticker or photos. That content is then visible through WallaMe's AR viewer to any other user passing by. (Messages can be made private and visible only to specific users if preferred.) Think of it as digital graffiti or a way of sending underground messages to people in the same proximity. Pretty cool.

Social media heads in a more immersive direction

When Facebook first started, it was a place for us to chat with friends, share life news and perfect the art of humble-bragging. Nowadays, it's more about businesses competing for customers and market share, and a place for interested parties to share (often questionable) news and information. There's no doubt Facebook is still a hugely influential force in our lives – and our elections – but is it really the (virtual) place we go to hang out and connect with our friends? For many people, not so much. Which begs the question, where does the future lie for Facebook?

The answer may lie in VR, or "social VR." Using VR, social media has the potential to become much more immersive and allow users to interact with each other in new, exciting ways. That's the idea behind Facebook Horizon, which launched its public beta version just as I was writing this chapter. Facebook has invested heavily in

VR, purchasing the Oculus VR brand (see Chapter 2) and investing in VR hardware – even releasing an impressive prototype of a VR headset that looks like sunglasses.[iv] Horizon, the company's latest big step into VR, is a virtual reality social networking platform where users can meet up, hang out and play games with friends. Using one of Facebook's Oculus VR headsets, you enter the platform as a floating avatar, and you and your friends have the ability to create entirely new worlds and hangout spots of your own design, as well as play games and activities built using the platform's internal tools.

This reminds me so much of the movie *Ready Player One*, where the virtual world becomes far more enticing than the real one, you can earn coins and status in the virtual world, and even become a completely different person if you want to. Read more about this merging of the real and digital worlds for entertainment in Chapter 8.

Improving the dating experience

As well as social media, VR could revolutionize the experience of dating or maintaining a long-distance relationship. The idea is couples can share virtual environments and experiences, even if they're thousands of miles apart. In other words, with both people using VR headsets, a couple could enjoy a date in Paris, watch the sunset on a beach in Thailand, or even explore Everest Base Camp.

Clearly, chatting on Skype or other video platforms is a far better way to connect with someone romantically than just talking on the phone, so it makes sense that VR could enhance this connection even further, and provide a low-cost, low-risk way to meet new people and explore potential love matches (or, maintain the spark in a long-distance relationship). The League dating app has used VR to provide a "virtual blind date party,"[v] and it's reasonable to expect other dating apps will trial their own VR experiences in the near future.

Looking further ahead, the virtual dating experience could become much more immersive, to the point where you can "feel" the other person and smell their cologne. One report by eHarmony and Imperial College Business School predicted we'll be able to go on full-sensory virtual dates by 2040.[vi]

Bringing news, history and world issues to life

A number of news organizations are beginning to experiment with XR technologies, particularly AR, as a way to enhance news coverage. For example, *The New York Times* app allows readers to view selected stories with additional AR features. Likewise, fashion magazine *W* has used AR to deliver interactive behind-the-scenes content.

XR technologies also provide a wonderful way to bring moments and artefacts from history to life. Take Apollo 11, the spacecraft used in the first lunar landing, as an example. While you can see the spacecraft on display at the Smithsonian's National Air and Space Museum, it's encased in a protective plastic shell and visitors have no sense of what it was really like to be inside. To remedy this, the Smithsonian collaborated with Autodesk to create a 3D experience that re-creates the inside of Apollo 11 in fine detail. There's also an Apollo's Moon Shot AR app that places users on the surface of the Moon.

Then there's technology's ability to bring important social issues to life. At the time of writing this book, Black Lives Matter protests were taking place across the world, yet many of us have no idea what it's really like to live life as a person of color. I certainly don't. That's where VR can help, by giving us the chance to explore the world from another person's point of view. An example of this is the VR film *Traveling While Black*, an Emmy-nominated documentary that gives viewers an immersive historical experience of the dangers of being black in America.[vii] Watching it is a transformative experience, and I hope to see more XR projects that aim to address inequalities.

And, finally, some other examples from everyday life

Let me finish this section on everyday XR with a few random examples. I've mentioned the use of heads-up displays (HUDs) already in the book. While writing this chapter, Mercedes unveiled a sneak peak of a huge HUD screen coming to the 2021 S-Class, which uses AR to show navigation instructions.[viii] This is just the beginning of AR in automobiles, and we can expect more car manufacturers to incorporate HUD screens in their vehicles from now on.

Another interesting example comes from home improvement chain Lowe's. The firm's Measured by Lowe's app uses AR to turn your smartphone into a tape measure. Plus, there's a separate Lowe's app called Envisioned that allows customers to superimpose 3D images of Lowe's products into their homes. The app scans your surroundings, so virtual objects can be dropped in in their actual size and be dragged wherever you like in the room. Read more examples from the world of retail in Chapter 5.

And finally, XR technologies aren't just for humans. One Russian farm has experimented with giving cows specially adapted VR headsets that show a simulation of lush summer fields. (Russian winters being famously long and hard, and summers fleeting.) The VR simulation reportedly helped to reduce the cows' anxiety and improve "the overall emotional mood of the herd."[ix] A long-term study is planned to explore the effects in more detail, including whether it leads to increased milk production. If successful, it could be rolled out to other Russian farms. So, you see, even cows will soon be living in a virtual world . . ..

XR in the Workplace

Now let's turn to the workplace, where XR technologies – particularly VR – are making big waves.

Enhancing various recruitment processes

A range of recruitment processes can be enhanced through XR, from interviewing candidates to onboarding new hires. VR, in particular, is very useful for assessing candidates and giving prospective employees an idea of what it's really like to work for the company. Using VR, you can immerse the candidate in a highly realistic simulation of a particular role or function and see how they perform. Lloyds Banking Group is one early adopter of this sort of VR-based assessment. In a Lloyds VR assessment center, candidates put on VR goggles and are transported to a digital environment where they work through a range of workplace scenarios.[x]

Professional services company Accenture also uses VR to assess candidates – adopting a more creative, storytelling approach. Using a VR headset, candidates enter a virtual Egyptian crypt, where they are asked to solve a series of hieroglyphics. The exercise is designed to identify candidates who have the aptitude and potential to become software programmers. The company says this approach helps to foster diversity, since it assesses candidates based on potential, not academic skills on paper – as shown by the experience of a dyslexic candidate who shined in the hieroglyphics assessment in a way they hadn't with other organizations' traditional assessments.[xi] The company has also said this VR-based recruitment has really helped to differentiate them in the highly competitive graduate recruitment arena, attracting 20,000 graduate applications in half the time it would usually take.

That's a really important point – VR experiences can be an incredibly useful way to enhance your employer brand and attract candidates who might not otherwise consider your firm. For example, it's fair to say that food service company Compass Group lacks that household name awareness, yet it's a huge employer with over 500,000 employees. This lack of brand awareness can make attracting talented graduates a real challenge, so Compass created a VR experience for campus events that lets students take a virtual tour of the workplace and participate in a video interview.[xii]

And for candidates, VR provides a handy way to practice their interview skills and prepare for future interviews. This is the idea behind the Virtual Speech tool, which combines online tutorials with VR simulations to help candidates improve their skills and reduce their nerves in a safe, supportive environment.

(I should say that, although the use of XR in recruitment has primarily focused on VR, it's not the only option. New Zealand bank ASB has used AR to recruit corporate bankers. With the help of an AR app, candidates scan a recruitment pamphlet and bring the content to life. VR is far more common, however.)

What about once you've found the perfect hires? XR can also help you onboard them successfully, as IKEA, the world's largest furniture retailer, demonstrates. Retail has a fairly high employee turnover, and IKEA is no exception, so investing in proper onboarding and training is a must. Partnering with VR firm Virsabi, IKEA created a VR experience that enables employees to understand different job functions and experience a leadership role. The 360-degree VR video introduces colleagues to two real leaders in different IKEA environments, in order to demonstrate how IKEA's leaders are inspired by the company's eight key values, such as "togetherness" and "simplicity."[xiii] There's also a gamification element, where colleagues can interact with IKEA's key values as a game. Read more about the use of XR in training and education in Chapter 6.

Making data analysis easier and more immersive

VR also gives us many new and exciting ways to visualize and understand data, and generally make information much more immersive.

Take BadVR, whose tagline is "Step inside your data," as an example. I've written entire books on the importance of data in business operations, decision making and understanding customers, so I welcome anything that makes data easier to understand and, crucially, *act upon*. BadVR's mission is to use immersive technology as a scalable, fast way to visualize and analyze nearly any type of dataset, big or small, and turn it into actionable insights. So, rather than a series of impenetrable numbers and boring pie graphs, users can (using a VR headset) literally step into their data and see them in exciting new ways, such as floating objects or color-coded symbols.

In finance, Citi Bank has explored XR technology as a way to visualize its data, creating a "Holographic Workstation" for financial traders. Using Microsoft's HoloLens technology, Citi Bank created an integrated 2D and 3D system that enables traders to visualize financial data in real time and track trends, all using holograms.[xiv] The proof of concept is less immersive than something like BadVR, but as a way to make data more interactive and improve financial decision making, it's certainly a promising idea. It also makes sharing data easier, since users of the Holographic Workstation can share their interactive visualizations with others in real time using voice commands.

Improving virtual meetings

I've already mentioned in this book how, during the coronavirus pandemic, Zoom meetings quickly became the norm for many people. Personally, I believe increased remote working and virtual meetings are here to stay for most industries, so we can expect Zoom and its competitors to continue to grow in popularity – and to continually add new features designed to make our virtual meetings better.

How about a feature that smooths out your skin and minimizes the appearance of perfections? Already done. Zoom's "touch up my appearance" option is an AR filter that softens your appearance, smooths out your skin tone and generally makes you look more polished. In a time when many of us have spent more time indoors than normal, features like this help us look perkier and more refreshed than we probably feel in real life. Unfortunately, it doesn't filter out sweatpants, so you still have to dress appropriately for the occasion! But who knows whether a "touch up my outfit" feature will be available in future? What we do know is XR-related filters and features will become more and more common.

Transitioning to virtual trade shows

As a frequent business traveler, I'm well aware how the coronavirus pandemic utterly transformed business travel, perhaps forever. Meetings, training sessions, strategy sessions and conferences . . . a lot of events moved online with relative ease, prompting many of us to question whether we really need to go back to clocking up quite so many (road and air) miles in future.

But could major national trade shows and exhibitions move online quite so seamlessly? If the Thin Air trade show, the world's first truly digital trade show, is anything to go by, the answer is yes. Taking place in September 2020 and developed by outdoor gear review platform Gearmunk, the Thin Air virtual media show was an entirely new experience for the outdoor industry, designed to challenge traditional shows like Outdoor Retailer. The goal was to mimic the experience of a real-world trade show, right down to the finest details. So, after registering and creating their personalized avatars, attendees could wander through the convention center hallways, introduce themselves to other attendees, and even hear what other people were saying as they pass by. (Possible thanks to spatial 3D audio, which means if you pass between two people, you hear two voices through opposite sides of your speakers or headphones. Or if you turn your back on someone, their voice will get quieter, just as in real life.)

The goal, then, was clearly to replicate the important social aspects of attending such shows, as well as achieving the usual business benefits. There were booths, presentation halls and networking areas, just as in a regular show, and booths were staffed by avatars of brand representatives – all live and in real time.

Virtual shows like this are clearly far cheaper, more flexible and less environmentally damaging than in-person trade shows. Going forward, I think we'll see more trade shows and exhibitions move online, or at least transition to a blend of in-person and virtual events.

And why not, when online spaces are becoming increasingly realistic and immersive? For example, with tools like the Matterport Capture app, any real-life space can be scanned and turned into a 3D setting – meaning you could easily make a virtual 3D replica (a digital twin, if you like) of your office space or other environment, big or small. This could be used to create promotional videos, virtual walk-throughs or even VR experiences. All you need is a compatible camera (including an iPhone camera), and the Capture app scans and creates a dimensionally accurate digital replica. You can edit and customize the space (for example, blurring employees' faces), add additional details and then share it with others using the Matterport Showcase app. With Showcase, others can experience your space in walk-through mode, dollhouse view, floorplan mode and as an immersive VR experience for Cardboard and Oculus devices. Tools like this help to make simple VR experiences open to all types of users.

Key Takeaways

In this chapter, we've learned:

- XR is already widely used in everyday life, particularly in the form of AR-based apps, tools and filters. The latest smartphones and tablets come equipped with AR technology that, with the right apps, can superimpose fun, entertaining and informative content onto real life – from daft Snapchat filters to 3D Google search results for animals.

- And in the world of work, XR (especially VR) is being adopted in a wide range of functions, including recruitment, onboarding, training, data visualization, virtual meetings and even online trade shows.

As I alluded to in this chapter, one of the most common ways brands are using XR technologies is to boost customer engagement. Simple AR filters are a popular way to do this, but some brands have splashed out on much more impressive AR experiences. For example, Pepsi created an incredible AR display in a London bus shelter, which stunned commuters by overlaying eye-popping images – like a meteor crashing into the ground, or a tiger padding toward them – onto the real-life street in front of them. Pepsi's video of shocked, dumbfounded commuters is worth a watch. Not to be outdone, Uber installed an AR experience at a Zurich train station that immersed passers-by in adventures such as petting a tiger in the jungle. A video of people interacting with the experience has had more than a million YouTube views.[xv]

In the next chapter, I explore this notion of using XR to attract and engage customers in much more detail.

Endnotes

i. Live Texturing of Augmented Reality Characters from Colored Drawings; Disney Research; https://la.disneyresearch.com/publication/live-texturing-of-augmented-reality-characters/
ii. Facebook Shares Major Spark AR Studio Update; Facebook; https://developers.facebook.com/blog/post/2019/04/30/spark-ar-studio-update/

iii. What Does the Future Hold for Augmented Reality in Digital Marketing?; Rubix; https://rubixmarketing.uk/2018/04/06/augmented-reality-digital-marketing/

iv. Facebook's newest proof-of-concept VR headset looks like a pair of sunglasses; The Verge; https://www.theverge.com/2020/6/30/21308813/facebook-vr-sunglasses-research-proof-of-concept

v. "The League" is Hosting a VR Blind Date Party on Valentine's Day; Trendhunter; https://www.trendhunter.com/trends/vr-blind-date-party

vi. The future of dating: 2040; eHarmony; https://www.eharmony.co.uk/dating-advice/wp-content/uploads/2015/11/eHarmony.co_.uk-Imperial-College-Future-of-Dating-Report-20401.pdf

vii. *Traveling While Black:* Behind the eye-opening VR documentary on racism in America; The Guardian; https://www.theguardian.com/tv-and-radio/2019/sep/02/traveling-while-black-behind-the-eye-opening-vr-documentary-on-racism-in-america

viii. 2021 Mercedes S-Class Shows Huge Screen, HUD with Augmented Reality; Motor1; https://www.motor1.com/news/432657/2021-mercedes-s-class-w223-interior/

ix. Russian cows get VR headsets "to reduce anxiety"; BBC News; https://www.bbc.com/news/world-europe-50571010

x. These 3 Business Functions Could Be Transformed By VR; Forbes; https://www.forbes.com/sites/bernardmarr/2020/07/31/these-3-business-functions-could-be-transformed-by-vr/#2beb5df021b1

xi. How augmented reality is infiltrating the world of HR; People Management; https://www.peoplemanagement.co.uk/long-reads/articles/augmented-reality-infiltrating-world-hr

xii. These 3 Business Functions Could Be Transformed By VR; Forbes; https://www.forbes.com/sites/bernardmarr/2020/07/31/these-3-business-functions-could-be-transformed-by-vr/#2beb5df021b1

xiii. IKEA is using virtual reality for onboarding and training; Virsabi; https://virsabi.com/ikea-is-using-virtual-reality-for-onboarding-and-training/

xiv. HoloLens could get into finance with this VR workstation; Mashable; https://mashable.com/2016/03/30/hololens-finance-citi/?europe=true#e0GtQZbZysqT

xv. Augmented reality experience at Zurich main station, Uber; YouTube; https://www.youtube.com/watch?v=bCcvEVyAXQ0

5
CUSTOMER ENGAGEMENT AND RETAIL

How do you make retail better? How do you engage people and make the customer journey more interesting? How do you ease the customer's decision making and improve conversion rates? How do you reduce returns (the cost of which is a huge burden for retailers, especially online retailers)? These are the sorts of questions businesses grapple with constantly.

XR technologies can help to smooth out the wrinkles in the buying journey, thus providing some solutions to these critical questions. For one thing, augmented reality allows customers to see and even try things out before they buy – for example, digitally projecting a new shade of nail polish onto their nails via an AR app, or digitally manifesting the latest gadget on their desk so they can inspect it from all angles. Thus, AR provides a way for customers to personalize their shopping experience and visualize purchases in context. But XR can also help to provide more immersive, meaningful experiences for customers – such as using VR to tell a story about the brand's identity or the history of a product. XR is even providing opportunities for some cutting-edge B2C businesses to create new, digital-only products for customers to enjoy purely in a virtual sense – including digital-only clothes. All this, and more, is now possible thanks to XR.

When you think about it, retail is ripe for an overhaul. Particularly the online shopping experience. Take shopping for furniture online as an example. There's a reason so many people still shop for furniture in person rather than online – judging the size of something like a couch is really tricky online, and then there's the issue of visualizing how it'll look in the context of your home. And if, once the couch arrives, you discover it doesn't fit or doesn't blend with the rest of the room, returning it is a nightmare. Now, retailers are using AR to help customers digitally project furniture (or other home items, and even paint colors) into their homes, scaled to size, so customers can judge accurately whether an item is right for them. The same sort of thing is now possible with clothes, cosmetics, glasses, shoes, jewelry, tattoos . . . anything, really.

Research shows that the majority of customers are not only open to these tools – they may be more likely to purchase an item when they've experienced it through AR, even if they weren't originally intending to purchase that item.

- A 2019 study found that 57 percent of customers in the UK said they would definitely or probably use VR/AR applications that provided more information on products. For customers in the United States, that figure was 62 percent.[i]

- One study from 2016 found that 72 percent of shoppers bought something they had not planned to because of AR, and that 55 percent of shoppers said AR makes shopping fun.[ii]

- In the wake of COVID-19, when physical stores in many locations were closed, retailers that used AR enjoyed a 19 percent spike in customer engagement. What's more, conversion rates increase by 90 percent among customers that engage with AR versus those that don't.[iii]

In this chapter, we'll explore inspiring, real-world XR examples from a whole host of direct-to-consumer businesses, from tattoo artists to furniture and car manufacturers to cosmetic companies. The chapter is broken down into the broad ways in which organizations are already using XR, namely:

- Giving customers interesting and immersive new experiences

- Allowing customers to see and experience products in more detail

- Enabling customers to virtually trial a product before they buy

- Providing new opportunities to customize products

- Creating innovative new and digital-only products

These sections represent the most common current XR uses, but as the technology evolves, we can expect to see different uses and experiences that are much more impressive. For example, for now, most AR clothes shopping experiences are limited to viewing items on models or avatars representing various sizes. But in the future, I'll be able to create a digital twin of myself that accurately represents my own size and body shape – so when I shop for clothes online, I'll be able to transport my digital avatar to a virtual changing room and see what the clothes will *really* look like on me. In other words, within the next few years, we'll see the retail experience transform in ways that we couldn't have imagined 20 or even 10 years ago.

Giving Customers More Immersive Experiences

VR and AR are giving brands new ways to engage with customers, by immersing them in informative, entertaining experiences. Let's look at some examples of XR being used to deepen customer engagement.

Foot Locker

American footwear and sportswear retailer Foot Locker prides itself on celebrating youth and sneaker culture. So, when launching its LeBron 16 King Court Purple sneakers, it makes sense that Foot Locker used technology to turn the event into a fun experience for sneakerheads.

Forget camping outside a store for hours to get your hands on a must-have new product – as you might for the latest iPhone release. Foot Locker devised an AR-based scavenger hunt, using an AR app and geo-targeted clues, to lead sneaker fans in Los Angeles to a location where they could purchase an early, limited edition release of the coveted sneakers. The launch experience, created in partnership with design and innovation agency Firstborn, was a hit and the sneakers sold out within two hours, demonstrating how AR can be used to create hype and boost sales.

Burger King

Foot Locker isn't the only company using AR in a quirky, fun way. American multinational Burger King welcomes more than 11 million guests every day, making it the second-largest fast-food burger chain in the world. But with an increasingly crowded fast-food scene, Burger King leapt on the chance to "burn" its rivals . . . using AR.

The company created an AR-enabled feature for the Burger King app called "Burn That Ad," which encouraged users to earn a free Whopper by "burning" the ads of rival fast-food chains. How? All they had to do was point their smartphone at a rival billboard or magazine ad and, thanks to the AR app feature, watch that ad go up in flames (playing on Burger King's famous flame-grilled flavor). After burning to a crisp, the rival ad was replaced by a picture of a Whopper, with a link to claim a free burger at the nearest restaurant. The genius of this campaign, which was available in Brazil and produced in partnership with Brazilian creative agency DAVID, is that it turned the media investments of rival companies into BK ads!

Red Bull

Austrian company Red Bull has the highest market share of any energy drink in the world, but is also known for its high-profile sponsorship of sporting events and athletes – across traditional sports and some sports that might be considered, well, less traditional.

So, when Red Bull launched its first web-based AR experience, the company partnered with pro gamer and influencer Tyler "Ninja" Blevins, known for streaming games like Halo and Fortnite. The Win with Ninja campaign was entirely web-based, meaning fans didn't need to download an app to enjoy it. Thanks to a simple AR Snapchat lens, activated by tapping a button on the Red Bull website, fans could have an interactive avatar of Ninja appear in their homes, where they could snap pictures of him and enter into a raffle to win a real gaming session with the star. Red Bull partnered with German Agency "and dos Santos" to create the experience, which also featured a limited-edition Red Bull can sporting Ninja's likeness. For me, the fact that users didn't need to download an app made this campaign impressively accessible and shareable.

LEGO

Danish toy company LEGO may have been around since 1932, but it's certainly not afraid to embrace new technologies and new mediums, including movies, mobile apps and robotics. In 2019, the company launched eight AR-focused LEGO sets called Hidden Side. All featuring haunted buildings, the range is designed to combine real life and the virtual world. The idea is children build a model of a haunted house, then use a free interactive AR app to hunt and trap ghosts – with the app telling a wider story about children uncovering mysterious happenings in their hometown. Even without buying one of the sets, LEGO fans can still use the app to play a standalone game – although the experience is obviously better when you have the physical LEGO in front of you. And, with a new multiplayer feature, one child can act as the ghosthunter while up to three friends can join in as ghosts. What I like about this is it bridges real-life and virtual play, and deepens the experience children have when playing with real-life models.

Mercedes-Benz

Now for something a little more grown-up. Founded in 1926, German car manufacturer Mercedes-Benz is known for its luxury vehicles. So, when it came to promoting a new model, it's no surprise Mercedes opted to immerse customers in a luxurious-looking campaign.

Previously, Mercedes had created a sleek virtual experience of driving the SL model down California's dramatic Pacific Coast Highway. Then, to celebrate the launch of the new E-Class, the company created a beautiful 360-degree video of the car cruising through Lisbon and the Portuguese countryside. This shows how XR-related technology can be used to create immersive experiences that enhance the customer's perception of a brand – in this case, how owning a Mercedes could transport you to a new, luxurious, altogether more beautiful life. You know the old adage "Show, don't tell?" Thanks to XR, showing customers becomes a lot more powerful.

On the theme of "Show, don't tell," let's look at a few examples of how AR and VR has been enthusiastically adopted by alcohol brands and bars to visually delight customers and entertain customers.

One Aldwych Hotel

Located in Covent Garden, London, One Aldwych is a five-star boutique hotel with all the luxury touches you'd expect, including a striking lobby bar. And it's here that a unique, limited-edition "VR cocktail" was created in 2017.

In case you're wondering, the cocktail itself was real. Named Origin, the drink was a blend of 12-year-old whisky, cherry liquor, cherry puree, grapefruit juice, chocolate bitters and champagne. The VR was more of a fancy garnish. Customers choosing the cocktail were first given a VR headset, which transported them, virtually, to the Dalmore distillery in the Scottish Highlands (Dalmore Whisky being the signature ingredient in the cocktail). Designed to showcase the origin of the drink, the VR experience flew customers to the distillery where the whisky is aged, then to the fields of barley and water sources that made the whisky. After soaring over the beautiful Highlands, customers then floated (again, virtually) back to the hotel's bar, where the film concluded with a bartender making the drink. On removing the goggles, customers would then see the drink in front of them, presented in exactly the same way. The bar sold 30 cocktails on its debut night, and reported a distinct domino effect – meaning, as customers saw other people enjoying the VR-infused drink, they wanted one for themselves. I particularly like how this experience showcases the whisky's origin story, making it a neat way to engage customers with a brand's history and values.

Miller Lite

Particularly in America, it's common for brands to create special campaigns for a variety of holidays. Now, these campaigns are taking on a new, more immersive twist, as shown by US-based beer brand Miller Lite – the original low-calorie, low-carb beer – and its St. Patrick's AR experience.

As a web-based AR experience, fans of the beer brand didn't need to download a dedicated app – they just had to visit a special page on the Miller Lite website on their mobile, and have a can of Miller Lite handy. On scanning the Miller Lite crest on the physical can, an AR leprechaun leaped out of the can into their live environment, entertaining users with tricks such as pulling a can of beer out of its beard. Crucially, users could snap pictures of the leprechaun and share them with friends on social media – thus, helping to raise brand awareness and trigger greater engagement with the campaign.

Related to this, AR-enabled labels and packaging are fast becoming a popular way for brands to share more information with their customers, as these final two examples demonstrate.

Bombay Sapphire

Premium gin brand Bombay Sapphire is known for its smooth taste and distinctive blue bottles. Thanks to AR, those bottles – or, more specifically, the labels on those bottles – became a lot more interactive.

Partnering with the popular app Shazam and AR company ZAP-PAR, Bombay Sapphire created a unique AR experience for customers that used the bottle label as a way to activate more information. When scanned using the Shazam app, beautiful botanicals appear to "grow" out from the label, representing the essence and character of the famous gin. Then, customers can tap the screen and get exclusive video content showcasing a range of different cocktail recipes – all featuring Bombay Sapphire gin, of course.

What's powerful about this example is it shows how pretty much any physical item can be used to deliver all kinds of additional digital content, including video, animation and textual information.

Living Wine Labels app

Not to be outdone by gin, wine labels are also getting in on the AR-enabled, interactive label trend. The Living Wine Labels app turns the labels on wine bottles into portals to interesting and informative AR content, all about wine.

Created by Treasury Wine Estates, one of the world's largest wine companies, and Unity Technologies, the free app allows wine lovers to enhance their experience of their favorite wine by accessing AR content, such as the history of the vineyard that particular wine came from, or tasting notes to help enjoy the wine. In one unique example, lovers of The Walking Dead wine (yes, there is such a thing) could use the app to bring Sheriff Rick from *The Walking Dead* TV show into their living room.

Bringing Products to Life in Greater Detail

AR and VR have enormous potential to ease buyers' decision-making processes by letting them see and experience products in greater detail before they buy. Indeed, many brands are investing in XR-enabled tools to help their customers get a good look at products from the comfort of their own homes.

Apple

Household name Apple is one of the world's most recognizable brands. For many, Apple products inspire almost fanatical levels of brand loyalty among customers, and it's not uncommon to see customers camping in long lines outside stores to get their hands on the newest iPhone or iPad release.

But that hasn't stopped Apple leveraging AR to help customers get a better look at products before they buy. To show off the iPhone

11 series, the company created a web-based AR demo (accessed via the Apple website on a mobile device) that let customers view and manipulate the new devices from their screen. Users could spin the iPhone around to view it from multiple angles, then switch to an AR mode to view it as a digital representation in the real world – giving fans a chance to see how it looks in real life and its actual size. They did a similar thing for the new iPad Pro, allowing customers to digitally place the new iPad on their desk or table and spin it around to view from different angles. Particularly when people are unable to visit a store to view a product in real life (for example, when there's a global pandemic under way), features like this allow customers to get a better feel for products.

Asos

Founded in 2000, British online fashion retailer Asos is a relatively young brand with a huge following among young adults. So, it makes sense that Asos would embrace AR as a way to enhance the customer's buying journey.

In 2020, Asos introduced an AR feature on its product pages that allows shoppers to see simulated views of models wearing products. Developed in partnership with AR and AI specialists Zeekit, the See My Fit feature is designed to help customers better gauge the size, cut and fit of garments – by showing the same items on multiple models of different heights and sizes. Because the images are simulated, the models aren't actually wearing the clothing, but as a way to gauge how clothes look on different body shapes (rather than guessing in your head), it's bound to prove useful for shoppers. And with poor sizing or fit being one of the main reasons customers return clothing bought online, AR-augmented product listings could help retailers like Asos cut returns.

Gap

Gap unveiled a similar pilot app back in 2017. Founded in 1969, American retailer Gap Inc. is one of the world's most recognizable apparel and accessories brands. Its signature style may be classic and comfortable, but Gap isn't afraid to push the envelope when it comes to the shopping experience. The company's pilot app, called DressingRoom by Gap, used avatars of different body shapes to show how Gap products fit different body types.

Sotheby's and virtual house tours

From buying clothes to buying homes, VR could help transform real estate sales by creating digital walkthroughs of property. The idea is a potential buyer can view the tour on any device from the comfort of their home and get a perfectly accurate virtual walkthrough that is indistinguishable from a real-life walkthrough. Alongside typical floorplans and still images, VR tours help buyers better understand a space – and with additional features like tasteful background music, can be used to create a specific ambiance.

VR property tours can be particularly useful for high-end buyers, which is the idea behind Sotheby's VR tours for luxury homes. Sotheby's International Realty arm has been selling luxury properties across more than 70 countries since 1976. But when your buyer is located on the other side of the world, the likelihood of them popping over for a quick property tour is pretty slim. So Sotheby's created VR tours that let viewers experience homes through a mobile device and compatible VR headset.

Obviously, virtual house tours will not replace in-person house tours. Rather, they enhance the real estate sales process by giving buyers detailed first viewings in an easier, less time-consuming way – the real estate agent only has to do physical viewings with serious potential buyers, and buyers don't have to waste their time traipsing around unsuitable properties.

Christie's

While we're on the subject of high-end auction houses, Christie's is also embracing XR technology. Founded in London in 1766, Christie's specializes in premier auctions and private sales of art, antiques, jewelry and more. Now, the company is using AR to enhance its fine art auctions, by allowing users to view paintings, drawings and prints in AR via the Christie's app. In other words, you can digitally hang a Rothko, Picasso, Monet and more on the wall of your home to see how it looks before you splash out some real-life cash.

Not many of us have the funds to purchase a masterpiece, but the idea of virtually viewing products in your real-life space is catching on right across the retail industry, particularly in homewares, furniture and home improvement.

Home Depot

Home Depot is the world's largest home improvement retailer, with almost 400,000 employees and 2,200 stores across the United States, Canada and Mexico. Now, customers browsing in Safari on their Apple devices can be treated to additional pop-up AR content called AR Quick Look – a feature that Apple has rolled out for many retailers and brands.

But Home Depot has also invested in AR technology of its own, with its Project Color app. The idea is to help shoppers find the exact right shade for their space and try out different colors in their rooms before they buy. In other words, you simply point your camera at a wall or room in your home, select a shade of paint in the app, and the space is transformed before your eyes with a true rendering of what you would see on your walls.

Finding the perfect shade can be a real bugbear for customers – I can't be the only person who's previously chosen a shade based on those little card swatches, bought the paint, slapped it on the walls at home only to discover it's not right for that room's proportions and light. Tools like this are therefore a godsend because they help to remove common pain points for customers, and save them wasting precious money and time on unsuitable products.

IKEA

Choosing the right shade of paint isn't the only tricky decision facing consumers. Investing in big-ticket furniture items also tends to create headaches and indecision. How will it look in the room? Will it match our other pieces? Will it even fit?

IKEA's app helps to answer these questions. Global phenomenon IKEA was founded in Sweden in 1943 and has since spread to more than 50 countries. The company itself estimates that almost every fourth person on the planet comes into contact with IKEA in some way, which is pretty impressive for a furniture retailer. The IKEA Place app lets customers digitally render Ikea furniture in their home to see whether it's right before they buy. Users scan the room in question to create an accurate representation, then they can place IKEA furniture and objects in the digital image of that room. IKEA promises that the more than 3,000 items available in the app are true to scale, with up to 98 percent accuracy, so you can be pretty sure that couch really will fit in that space. The company also claims the AR technology is precise enough to be able to see texture of fabrics and the interplay of light and shadows on furniture.

What we can learn from this is that retail is heading in a direction that leverages technology to ease the customer journey and make it easier to hit the "buy now" button. In fact, more and more retailers have been investing in their own AR apps and features.

Target

American retail corporation Target has enormous reach in the United States – the chain has stores in all 50 states, and it is estimated that more than 75 percent of the US population lives within 10 miles of a Target store.

Like IKEA, Target offers shoppers a way to "see" products in their homes before they buy. The See It in Your Space feature on the Target app uses AR to digitally overlay furniture and other pieces onto the user's real-life space.

Target has also created AR features in Snapchat – such as the Halloween-themed experience that let users see how they'd look in three different Target costumes (a witch, mermaid and a unicorn, in case you're curious).

Wayfair

American furniture and homeware retailer Wayfair was founded in 2002 initially to sell stereo racks and stands online. Since then, the e-commerce retailer has grown to one of the world's most popular online destinations for home goods, with more than 18 million products.

As an online retailer, Wayfair doesn't have the luxury of showrooms or retail outlets to let customers view their products in real life. But thanks to the AR-enabled View in Room 3D feature in its mobile app, Wayfair customers can easily visualize furniture and décor in their own home. It projects furniture or décor in 3D at full scale, so that shoppers can see whether it fits in the space and how it will look in the room. And, importantly, because the image is 3D, shoppers can walk around the digital object to view it from different sides. This (literally) brings a new dimension to superimposing products into a real-life space and allows customers to experience the product in much more detail before they decide whether it's right for them.

Amazon

E-commerce pioneer Amazon is no stranger to blazing a trail in online shopping. And in 2019, the company launched Showroom, a virtual showroom that lets users place furniture in a virtual living room, customize the room's decor and purchase the items. You essentially create your own personalized living room scene by choosing different wall and floor colors, then add in products such as chairs, coffee tables, rugs, and even art. Then, when you've filled the room with all your favorite products, you can transfer everything in the room to your shopping cart. The feature is available on Amazon's website and in its mobile app.

Toyota

Even car manufacturers are turning to AR to help customers see and experience products in more detail. Japanese multinational automotive manufacturer Toyota is one of the world's largest manufacturers. In 2019, Toyota GB, the company responsible for sales and aftercare service in the UK, created an AR iPad app that allowed consumers to "see inside" the C-HR hybrid model. Designed to be used in showrooms, trade shows and shopping centers, the app uses AR to overlay the inner workings of the car onto the physical exterior of vehicles – to show consumers how the hybrid technology works and the benefits it offers, as well as give information on key features, like the battery and fuel tank. What I like about this example is it shows how AR can be used to enhance the customer's physical, in-store experience. In theory, the idea could be applied to any physical product in any store – customers simple point their phone or tablet at a product and get additional information such as how it works and the benefits of choosing that particular product.

A New Age of "Try Before You Buy"

Allowing customers to see and experience products in more detail is one thing. But how about letting customers try them on for size, digitally? Being able to realistically try on a product at home or wherever makes the buying decision easier than ever – and helps to remove those annoying wrinkles from the online shopping experience. These days, we can digitally try on all kinds of things, including glasses, cosmetics, clothes, watches, and even tattoos. Largely this is achieved through AR apps or AR web experiences, but we also now have smart mirrors that combine AR, artificial intelligence (AI) and gesture recognition technology to digitally adjust your image to create a very realistic augmented reflection.

This combination of XR technologies and AI is particularly powerful. For example, an AR app may let you try out different hairstyles or hair colors to see how they look on you, before you take the plunge. Combine this with AI, and you have an app that learns what sort of hairstyles and colors you're interested in, analyses your face shape and skin tone, and makes smart recommendations on what sort of style and shade may suit you best.

Let's see how brands are taking "try before you buy" to a whole new level.

InkHunter

Are you sure you really want to get that giant dragon tattoo on your arm? Really? Why not see how it'd look in real life before you have it permanently etched on your skin? That's the idea behind San

Francisco-based InkHunter. Founded in 2014, InkHunter provides a mobile app for people who fancy a tattoo but want to be sure it's right for them first.

Using AR, the app projects any tattoo design onto your body in real time. So, you can choose from a gallery of designs, or upload your own tat idea, then size and position the design on your body. As you look through your smartphone screen, you see a new, tattooed version of yourself. If you like what you see, you can snap a photo, then share it with your tattoo artist, or with friends to get a second opinion. There's even a blur effect, which presumably shows what your tattoo might look like after several years, when the lines are no longer as crisp. I particularly like how this tool not only shows you what the tattoo would look like now but also how it'll age on your skin – an idea that could be useful for other products, too (such as visualizing how an investment piece, like a luxury leather couch, will soften and age).

Skin Motion Soundwave Tattoos

Not sure about that dragon tattoo after all? How about getting a minute of sound inked on your body? Invented by a Los Angeles-based tattoo artist, Skin Motion's Soundwave Tattoos allow you to turn an audio recording of your loved one's voice, your favorite song or another sound into a beautiful soundwave design – that can be played back on your mobile, via the Skin Motion app. You upload your audio recording to the app, which generates a soundwave pattern, then take that design to a certified tattoo artist from the Skin Motion Artist Network who will ink it onto your body. For the playback to work, the soundwave has to be inked on a flat part of the body, like an arm, and can't curve around body parts. Within 24 hours of having the tattoo done, you can then (for a subscription fee) play it back through the app, simply by pointing your phone camera at the tattoo. While it's not strictly a "try before you buy" example, this shows how the whole customer journey can be enhanced through XR technologies – first, allowing customers to try a product before they buy, then allowing them to experience it in exciting new ways long after purchasing.

Warby Parker

Founded in 2010, New York–based Warby Parker is an online retailer of prescription glasses and sunglasses. Traditionally, choosing a new set of frames is a painful process; what looks good on the shelf doesn't look right on you, and the only way to know what *does* look good is to spend ages in store, trying on style after style, as a store assistant awkwardly follows you around. I know this because my teenage daughter is a glasses-wearer. She wants the latest trendy frames but doesn't particularly want to spend her Saturday morning in a store – and, frankly, neither does the rest of the family. So, you can imagine our relief at the launch of Warby Parker's award-winning AR-enabled app, which lets customers try on eyewear via their phone.

It makes sense that Warby Parker, a direct-to-consumer (mainly) online retailer, would be leading the way in this field, when the traditional in-store method of trying on frames isn't an option. The company used to get around this by letting customers choose five frames from the website, then mailing them to the customer to try on at home for free before committing to a purchase – which, while okay, was hardly the speediest way to choose glasses.

Warby Parker overhauled this process with its AR app, which uses face-mapping technology to show customers what their face would look like wearing different frames. The detail in this app is impressive – the virtual glasses stay in the right place even as your turn and tilt your head, and the app shows how light plays on different frames and metal details. For me, this shows how high-quality AR experiences can drastically improve the customer experience – and turn a potential customer into a devotee.

Nike

Buying shoes online is another shopping experience fraught with difficulty, particularly if you're between sizes or come up different sizes in different brands. American multinational Nike is known for its on-trend footwear and apparel – and, anecdotally, for sizing small across many styles. So, if you're a size 11 in most brands, you may find you're a size 12 in a Nike style. Or you may be a size 12 in Nike's Air Max designs, but an 11.5 in the Epic Reacts style. The only way to be sure is to head to a store and try them on for yourself. Or, at least, that used to be the only way to be sure.

In 2019, Nike updated its app with a new AR tool that measures customers' feet so they can buy sneakers that fit properly. Using the Nike Fit app feature, all you do is select which style of sneaker you're interested in, then stand next to a wall and point your camera at your feet. Once the app recognizes your feet, it scans them, measures them to within a two-millimeter accuracy, and tells you your ideal shoe size for the style you've chosen. Nike is so confident of its measuring app, it intends to use the technology in stores as well,[iv] showing again how XR technologies can enhance the in-store as well as online shopping experience.

Another benefit of tools like this is they make parents' lives easier. Young children especially are constantly outgrowing their shoes, so the ability to accurately measure your kids' feet at home and order the correct size shoe online, without dragging them to a store, is hugely attractive.

Watches of Switzerland and Grand Seiko

Revered by watch collectors, Japanese watchmaker Grand Seiko is one of the few remaining watch companies that makes virtually every part of its timepieces in-house. Now, I absolutely love classic timepieces, but even I would think twice before splashing out on an expensive luxury watch. To convert buyers, Grand Seiko and leading luxury watch specialist retailer Watches of Switzerland created an AR Instagram filter that lets people project watches onto their wrist and take a picture to share with friends on the platform. In other words, you can see what a beautiful watch would look like on your wrist, without having to visit a boutique. You can also rotate the watch with your fingers for a view of its back. The AR filter was created using Facebook's Spark AR Studio software.

WatchBox

On the subject of watches, global ecommerce platform WatchBox – which specializes in luxury pre-owned watches – has created its own app-based AR feature that lets customers try on luxury watches before they commit to buying. The watch digitally appears on the user's wrist in an accurate representation of the watch's real-life size, shape and dimensions. Users can also take pictures of the watch on their wrist to share with friends, demonstrating how the ability to share virtual try-ons with friends and family (for that all-important second opinion) is becoming an increasingly popular tool.

FaceCake

If you've ever bought jewelry online, you'll know how hard it is to gauge actual size and fit based on an image. This is where AR-based try-on apps excel, as California startup FaceCake Marketing Technologies knows all too well. The software and marketing company is a pioneer in shopping platform innovation and using AR to overcome traditional shopping barriers.

One of FaceCake's many AR innovations is Dangle, an AR tool that lets users try on earrings virtually. With lifelike movement and sizing relative to the user's face, shoppers get an accurate picture of how pieces will look on them. Many jewelry retailers have followed suit and created their own virtual try-on apps, including Indian online and physical jewelry retailer CaratLane, whose own try-on app features rings, earrings and necklaces.

Tenth Street Hats

Not a hat person? Perhaps an AR solution that lets you try on hats at home could convert you into a loyal hat fan. That's part of the fun behind family-owned retailer Tenth Street Hats' AR experience. The company – which was established in 1921 and is named after the location of its original warehouse in Oakland, California – partnered with AR solutions specialist Vertebrae to create a platform where users can try on a range of hats, view them from any angle and snap pictures of their favorite looks.

To measure its success, Tenth Street Hats monitored whether pages that funneled customers to the AR experience led to more conversions than regular product pages. The result was an impressive 33 percent increase in conversions and a 74% boost in engagement,[v] showing that AR doesn't just provide a fun experience for customers – it can have a demonstrable impact on customer engagement and the bottom line.

Sally Hansen

Established in 1946 by Sally Hansen, the eponymous American beauty brand has become a household name in the field of nailcare. In 2020, the brand launched a new AR-enabled Snapchat filter that lets customers try out nail polish shades virtually. Crucially, the lens works on every nail shape and skin tone, ensuring the virtual shade perfectly matches what you would see in real life. What's really clever about this Snapchat feature is, once customers have found a shade they love, they can simply click a button in Snapchat and go straight to the page to buy that exact polish – making the customer journey easier than ever. (Or they can just snap a picture and pretend they've had a manicure.)

Sephora

French multinational beauty retailer Sephora has more than 2,000 stores around the world, selling a variety of prestige personal care and beauty products. The company has used AR as part of its successful digital transformation – a transformation that has seen Sephora become the world's number one specialty beauty retailer. Among the company's digital projects is Sephora Virtual Artist, an app that lets users try on thousands of shades of lipstick and other makeup products, virtually. There's also a Color Match AI feature that helps customers find the ideal shade for their skin tone. The company worked with AR beauty specialist ModiFace to develop the technology, which uses facial tracking technology to accurately measure and track facial features in real time to create a realistic view.

L'Oréal

For more than 110 years, L'Oréal has been at the leading edge of personal care, hair care and cosmetics. The company's Style My Hair app, which was created in partnership with ModiFace, allows customers to try out new hair colors instantly, at the touch of a button. The technology digitally colors hair strand by strand to create a fully dimensional, realistic look – that can be easily undone if you don't like the end result. As well as customers using the app at home, it can also be used in salon consultations, to help stylists show clients what to expect in a more effective, realistic way – demonstrating how AR can be used not just to increase customer engagement and aid the buying process but also to improve business processes and interactions with customers. For L'Oréal, the benefits are clear – the company reported that using AR to showcase products triples conversion rates.[vi]

Why Not Customize Before You Buy?

The next step up from "try before you buy" is to customize and configure products, virtually, before you buy. This gives customers a unique chance to try out different iterations and find the setup that's just right for them.

In theory, in the future this could be applied to pretty much any kind of purchase, particularly as personalization, 3D printing and on-demand manufacturing become more accessible and affordable. In other words, you could customize the perfect pair of sneakers, see how they look virtually, get them sized accurately, and then have those shoes custom made to your exact specifications. That's the idea. For now, though, practical examples of XR-enabled customization come mainly from the fields of automobiles and real estate. Let's take a look.

Porsche

German automobile manufacturer Porsche specializes in high-performance sports cars, sport utility vehicles and sedans. In 2019, the company unveiled a new 3D AR app that lets Porsche fans digitally customize and create their perfect Porsche. Called AR Visualizer, the app generates a 3D photorealistic image of your ideal car right before your eyes. So, using the previously existing Porsche Car Configurator features, you can customize a model by choosing your perfect color, alloy wheels, and so on, then project a 3D image of that dream car onto your driveway, in your garage, or even in your living room. There's also a "highlight" feature, an X-ray-style cutout view of your chosen car that shows off the hidden technical details. It's easy to see how tools like this increase customer engagement, since you not only get to play around with different configurations – you then get to see how the product would look in the context of your real life.

BMW

With more than 125,000 employees and 31 production locations across 15 countries, BMW is one of the world's most successful manufacturers of premium cars and motorcycles. In 2017, the company unveiled an AR tool called iVisualizer. Like Porsche's AR offering, BMW's iVisualizer tool lets users see, customize and materialize a full-size BMW on their driveway, or wherever. Given that luxury car manufacturers are investing in such technology, it's reasonable to expect virtual customization tools will soon filter down to the rest of the automobile industry.

Urbanist Architecture

London architecture firm Urbanist Architecture specializes in residential renovations, extensions and new build homes. Undertaking a big renovation or extension project is never an easy decision. (All that expense, disruption and mess . . .) But when you can accurately plan, view and customize your new home interior, before you knock a single wall down, it makes taking the plunge that little bit easier. That's why Urbanist Architecture uses VR to help clients experience a design before it is built.

Using 4D VR technology, the company creates a realistic digital representation of the new design, including space planning, materials, all internal and external fixtures and fittings, furniture, and more. In other words, clients can walk around their new build house, extension or renovated home before it has actually been completed – which gives clients even greater input into the design stage, since they can check out all the finer details and transform the space exactly to their liking before the project gets under way. Urbanist Architecture is one of the first architects in London to deploy this technology, but I'm certain they won't be the last. I can see how VR tools like this will help construction projects run more smoothly, because (in theory, at least) it'll be far easier to make key design decisions at an earlier stage.

Creating Exciting New and Digital-Only Products for Customers

At the very cutting edge, XR technologies also bring unique opportunities to create entirely new and digital-only products for customers that are designed to enhance their virtual lives. I predict this will become a considerable growth area in future as consumers – particularly younger consumers – spend more and more time in digital spaces. But what do I mean by digital-only products? Let's say you regularly hang out with friends in a social VR space (see Facebook Horizon, Chapter 4). In the future, it might be possible to buy digital copies of your favorite artwork and interior design pieces to furnish your virtual hangout zone. Sound far-fetched? I don't think so, not when you consider that people already spend money on digital clothes that they can't wear in real life. The world's first piece of digital couture, created by The Fabricant, sold for $9,500 in 2019,[vii] and some gamers regularly spend money on virtual outfits for their gaming characters.

Fortnite "skins"

Fortnite is an online multiplayer video game that has dominated the gaming world since it was released in 2017. Millions of fans play the battle royal–style game daily across various different platforms. The game is free to play, but the Fortnite franchise makes its money through microtransactions made in the game – meaning players can purchase (with their real-life money) special outfits for their character in the game. Virtual swag, if you like.

These virtual outfits, known as "skins," offer no competitive advantage whatsoever. So why do players buy them? Largely, gamers are looking to personalize their gaming experience, stand out from the default character appearance, and show that they're cooler than other players (just like regular fashion, skins come in and out of fashion, and young players report being shamed for not wearing the latest skins).[viii] This means, although most skins only cost around $20, players tend to buy new skins to keep up with the latest releases – like the Ninja skin, released in 2020, modeled on pro gamer Tyler "Ninja" Blevins. An impressive 70 percent of Fortnite players have made in-game purchases, spending $85 each on average – and with 250 million global players, that really adds up.[ix]

It's not unreasonable to expect this notion of digital outfits to extend beyond gaming – so that you can wear your favorite designer pieces or sneakers in the virtual world, as well as (or instead of) the real one. XR technologies make this very easy to achieve. In fact, one fashion label already specializes in virtual clothing.

Happy99

The fast fashion industry is frequently criticized for harming the environment, paying workers poorly and increasing waste. Happy99, a fashion brand created by Nathalie Nguyen and Dominic Lopez, positions itself as the polar opposite to fast fashion and overconsumption – by creating digital-only shoes that don't exist in real life. Happy99's signature style could be described as a mixture of cyberpunk, neon clubwear and futuristic space shoes. Big, bold and kind of crazy-looking, the shoes look like cartoon footwear – and you can buy them for a one-off, digitally augmented photo that you can share with your online followers. It makes sense in a way, since many of us are already using Instagram and Snapchat AR filters to augment our appearance. Why not throw on a digital outfit as well?

Carlings

Scandinavian retailer Carlings, best known for its jeans, released its first digital clothing collection in 2018. Designed to raise money for WaterAid, the collection of 19 genderless pieces cost between $10 and $30 each, and customers simply sent in their image to be digitally kitted out.

Since then, Carlings has released an AR t-shirt – a physical t-shirt that can be augmented with AR. In other words, the t-shirt appears almost blank in real life but, when viewed through a smart phone, it shows an animated design. At the time of writing, customers can choose between 20 different digital designs, each with a timely political or social message.

Clearly, clothing won't be entirely digital in future (although the image of a naked population walking around looking at each other's AR "clothes" through smartphone screens is pretty funny). But digital clothing provides a new way for brands to engage with fans. Particularly with Gen Z consumers who have grown up spending money on the likes of Fortnite skins, the idea of purchasing digital clothing to impress their Instagram followers is hardly far-fetched. Even beyond the world of fashion and the realm of Gen Z-ers, the potential to customize virtual spaces with digital products will tempt many. How about using AR to hang a Rembrandt on the wall behind you for your Zoom calls? Or getting your hair digitally styled for an important photo? In the future, anything could be possible. Companies would do well to remember that, and consider their audience's appetite for digital-only products.

Lessons We Can Learn from the Retail Industry

I hope these examples inspire retailers and brands to consider XR technologies as a way to raise brand awareness, deepen engagement and improve the customer experience. But where should you start, and what should you avoid?

Based on these examples, there are some clear lessons we can learn about using XR:

- Firstly, boosting engagement through XR is about *enhancing the customer experience* in some way – not showing off a flashy AR or VR experience for the sake of it. Take the Aldwych One VR cocktail as an example. Priced at £18 (around $23), it was hardly a cheap drink, but the addition of VR created a memorable experience that, at the very least, gave customers a cool story to share with friends (but, I'm sure, also inspired a deeper love of whisky and greater understanding of the care that goes into crafting quality whisky).

- Secondly, the technology also needs to be accurate. There's no point creating an AR tool that lets users trial a product in their home (or in the case of watches, on their wrist) if it's not accurately scaled to size. If it doesn't do an accurate job, it completely fails to make the customer's life – and their buying decision – easier, defeating the purpose of the whole exercise.

- Finally, as the sheer breadth of examples in this chapter shows, AR in particular is incredibly accessible for businesses of all shapes and sizes. There are a range of ready-made tools and licenses for developing AR apps and experiences, including Google's ARCore, ARKit by Apple and Spark AR Studio by Facebook. But it's also interesting that many brands create their AR apps or experiences by partnering with XR specialists – which is a great option if you don't have XR capabilities in-house. Depending on your in-house skills, the right partner for you may combine technical XR expertise with creative expertise.

Key Takeaways

In this chapter, we've learned:

- XR technologies, particularly AR, will play a key role in the future of retail and customer engagement.

- XR is currently being used by retailers to provide more interesting and immersive experiences for customers; allowing customers to see and experience products in more detail; and even to trial them out, virtually, before they buy. All of this helps to engage people and improve the customer journey, as well as improve conversion rates and reduce returns.

- At the more cutting-edge end of retail, XR technologies also provide new opportunities to customize products, and even to create innovative, digital-only products.

This chapter has been all about connecting with consumers, but XR is also being used to improve internal business processes, particularly when it comes to training colleagues. In the next chapter, I explore how XR is transforming the world of on-the-job training and the education sector at large.

Endnotes

i. The Future of Shopping: connected, virtual and augmented; Periscope by McKinsey; https://www.periscope-solutions.com/download.aspx?fileID=3600

ii. New Study Explores the Impact of Augmented Reality on Retail; Business Wire; https://www.businesswire.com/news/home/20161018005039/en/New-Study-Explores-Impact-Augmented-Reality-Retail

iii. Why retailers should embrace augmented reality in the wake of COVID-19; Retail Customer Experience; https://www.retailcustomerexperience.com/articles/why-retailers-should-embrace-augmented-reality-in-the-wake-of-covid-19/

iv. Nike's new app uses AR to measure your feet to sell you sneakers that fit; The verge; https://www.theverge.com/2019/5/9/18538101/nike-fit-new-app-ar-measure-feet-shoe-size-online-order-augmented-reality

v. Tenth Street Hats sees 74.3% engagement jump with shoppable AR try-ons; Mobile Marketer; https://www.mobilemarketer.com/news/tenth-street-hats-sees-743-engagement-jump-with-shoppable-ar-try-ons/541947/

vi. Conversion rates triple when L'Oréal uses AR tech to showcase products; The Drum; https://www.thedrum.com/news/2019/07/02/conversion-rates-triple-when-l-or-al-uses-ar-tech-showcase-products

vii. World's First Digital Only Blockchain Clothing sells for $9,500; Forbes; https://www.forbes.com/sites/brookerobertsislam/2019/05/14/worlds-first-digital-only-blockchain-clothing-sells-for-9500/

viii. Kids who play Fortnite say they get bullied and shamed if they can't afford paid skins, according to a damning report on gaming habits; Business Insider; https://www.businessinsider.com/kids-feel-poor-if-they-dont-buy-custom-fortnite-skins-2019-10

ix. Fortnite Usage and Revenue Statistics (2020); Business of Apps; https://www.businessofapps.com/data/fortnite-statistics/

6
TRAINING AND EDUCATION

This was a fun chapter for me to write because there are so many incredible and inspiring XR examples from the world of education. But you might be wondering, why talk about education in what is primarily a business book? Well, for one thing, the EdTech (educational technology) sector is growing rapidly – at a reported annual rate of 17.9 percent – and is expected to reach USD 680 billion by 2027.[i] This means there are enticing opportunities for organizations to provide immersive, engaging learning opportunities – and that goes for personal improvement and lifelong learning as well as formal education in school and college environments. But there's also a move toward more immersive vocational and workplace training in many industries, so even if you aren't looking to create XR-enhanced educational experiences, still give this chapter a read. It could shape your internal training and development processes.

I believe education (of all kinds) is the key to success. It gives our children a firm academic base that helps them thrive later in life. And it gives all of us the opportunity to continually develop and improve, both in a personal and professional sense. But the act of learning isn't always easy (nor is teaching). Anything that makes learning – at any stage of life – more engaging and interesting, anything that makes it easier to absorb and remember information, is a benefit to society as a whole. In this tech-driven era, we have an unbelievable opportunity to aid the transfer of knowledge and transform the world of education. For me, this chapter points to the beginnings of an education revolution.

I'm married to a teacher, so I have no intention of disparaging current teaching methods, but it's fair to say that a lot of teaching is based on presenting students with facts. Successful students tend to be those who retain and recall facts easily, while students who struggle to process large amounts of information at one time can quickly become bored and disengage from the learning process, potentially disrupting those around them. Some people are visual learners, after all, and the ability to "see" a process, rather than read about it, is far more impactful for them.

XR technologies can help overcome these challenges and more by creating immersive, interactive worlds, where students can not only visualize concepts, they can also experience different times and places. This immersion can be achieved through virtual reality experiences that transport students away from the classroom. Or by augmented reality and mixed reality technology that brings concepts to life in the classroom setting, by overlaying graphics, text, animation and instructions onto the real world. So, rather than simply reading about something, students can experience it and even interact with it in exciting new ways. Students can learn by "living" experiences, in other worlds, which can provoke an important emotional connection to what they're learning, and make the information much more memorable. Evidence suggests that learning through experience in this way can increase the quality of learning and promote knowledge retention by as much as 75–90 percent.[ii]

XR technology can also open up educational experiences to a wider range of students – such as transporting students to different places and cultures, without the need for expensive field trips. What's more,

XR technologies enable students to learn by doing, through realistic simulations of tasks and assessments. As well as in formal education settings, this can be particularly valuable in corporate and vocational training – especially in high-risk jobs, as shown by some of the examples later in this chapter.

Over time, I believe the role of teacher or trainer is going to evolve from someone who delivers information or content to someone who *facilitates* content through a range of digital technologies. So, instead of presenting facts, teachers and trainers will create a learning environment in which students can truly explore topics in a more fun, engaging way – one that makes it easier to acquire, process and recall information.

As the examples in this chapter show, this vision is already beginning to take shape. So let's delve into some current, real-world examples of VR and AR in education.

XR in Personal Learning and Development

I'll start with some brief examples that demonstrate how VR and AR are improving learning and education in everyday life.

Purina

Purina, an American subsidiary of Nestlé, produces pet food, treats and other pet care products. The company's mission is to enhance the lives of pets and the people who love them. To aid this mission, Purina created an AR web experience called the "Purina ONE® 28-Day Challenge" designed to help pet owners learn about healthy pets. First, through the magic of AR, a virtual pet comes to life in your room. Then, as the pet leaps around enjoying itself, you learn how to spot the signs of a healthy pet, and how feeding your four-legged friend

Purina for 28 days could enhance their health. Purina partnered with XR app specialists Zappar to create the experience.

This is one of the less well-known AR examples you'll find in this book, but for me it demonstrates how brands can use AR to educate their customers with more interactive, engaging tools.

VirtualSpeech

Even if you've never heard of glossophobia, chances are you know someone who suffers from it. It's the fear of public speaking. Up to 77 percent of people suffer from some sort of anxiety around public speaking, and symptoms can include sweating, dry mouth, increased heart rate and nausea.[iii]

Award-winning VR provider VirtualSpeech, which was founded in 2016, has created a VR tool to help people practice public speaking in a more immersive, realistic way. Whether you want to master speaking in front of a large audience, become more confident in networking settings, or simply deliver better pitches and presentations to smaller audiences, VirtualSpeech's VR e-learning courses can help. After donning a VR headset, you find yourself in front of a simulated audience (there are a range of audience sizes to choose from). You can then practice your speech or presentation (with your real presentation slides, if you want), get real-time feedback on your delivery, and monitor your improvement over time. The tool can be used by individuals or by corporations to train their teams, and there's even a "Live VR" option, which features a trainer in the virtual setting. VirtualSpeech says its VR learning solutions have been used by more than 300,000 people across 130 countries – which demonstrates the growing appetite for VR-enhanced training, potentially across all kinds of soft skills.

University of Virginia

Teachers have to be pretty confident public speakers. Managing a class (which is, after all, an audience) and delivering information in an engaging way is an important part of being a good teacher – and, perhaps ironically, is one of the hardest aspects to, well, teach. A team from the University of Virginia Curry School of Education and Human Development set out to change that.

They created a groundbreaking VR-based "classroom simulator" to help people become better teachers. Trainee teachers can test and sharpen their delivery skills and learn how to manage the behavior of students in the classroom – all in an immersive virtual setting, complete with avatars that act like real children. Crucially, trainees get instant feedback from faculty advisors to help them improve, whereas in a real-life classroom setting, feedback on live classroom performance can take hours or even days. I love how this helps teachers prepare for the real-world challenges of a teaching career, in a supportive environment. But, as we'll see next, this isn't the only way teachers – and their students – are benefiting from XR.

Making Learning More Immersive for Students

Bringing us nicely into the world of teaching and education, let's see how XR, particularly VR, is helping to deliver more immersive educational experiences for students.

Labster

There is a wide range of XR-enabled apps designed to improve science education, including experiences that take you inside the human body (more on that coming up later). Labster is one such platform that's dedicated to enhancing science education by letting students experiment with hyper-realistic, simulated lab equipment – allowing them to conduct experiments in a risk-free way. At the time of writing, the technology was being used by universities in Denmark.

One study on virtual learning shows that students who practice with VR headsets outperform students that learn only via a desktop.[iv] In the experiment, two sets of students – one group assigned VR headsets and the other desktops – were given access to science simulations on their respective devices. Later, when asked to conduct an experiment in a real lab, the students who had used the VR simulation performed significantly better than the other group. A similar study of high school classrooms also found that virtual simulations significantly enhance students' scientific knowledge.[v] This points to the incredible value of VR simulations in science education.

Islands High School and MEL Chemistry VR

Chemistry has a reputation for being incredibly difficult, but one company is on a mission to make chemistry lessons much more interesting. MEL Chemistry VR is a series of over 30 VR lessons and tests that are aligned with the chemistry curriculum. Each one is no more than seven minutes long, meaning they can easily be incorporated in the flow of a regular classroom lesson. For each VR lesson, students pop on a VR headset and learn chemistry concepts in an accessible, interactive way – such as how the particles of an atom bounce around at different speeds depending on their state.

When Cristal Steele was a science teacher at Islands High School, Georgia, she used MEL Chemistry VR in her lessons and found it helped to improve students' focus and comprehension.[vi] According to Steele, in our increasingly tech-driven world, it's vital students have access to contemporary education sources that help them experience what they're learning first-hand. As one of her students told her, feeling like they were "part of" the educational material, rather than just copying it down, made the information stick with them. Having well and truly caught the VR bug, Steele is now setting up a purpose-built VR and AR lab at her current school, Beach High School. For me, the fact that VR can help students better absorb and remember what they're learning is a really key takeaway – and, in theory, applies to all kinds of learning, at all stages of our lives.

West Coast University – Los Angeles

XR isn't just enhancing education for younger students and high schoolers. In 2018, WCU-LA began exploring the use of AR and MR to improve student learning at university level. Partnering with Microsoft, WCU-LA created an innovative, individualized learning experience using Microsoft's HoloLens technology. Designed for anatomy students, the technology allows students to isolate, enlarge, dissect or even walk inside the human body. So, for example, students could walk through a human eye to see its components or see what happens to the heart during a heart attack – things that students would struggle to learn from an anatomy textbook. The result? Students who used the HoloLens saw a full letter grade improvement and a 10 percent improvement in test scores, which is really impressive.[vii]

1943: Berlin Blitz

VR also gives us exciting new ways to bring history to life, as shown by this VR experience, which transports students to the events of World War II. Produced for the BBC by Immersive VR Education, 1943: Berlin Blitz recreates the events of the night of September 3, 1943. Specifically, it portrays the true story of a Lancaster Bomber on a mission to Berlin. What was special about this flight is, alongside the regular seven-man crew, the plane was also carrying BBC reporter Wynford Vaughan-Thomas and his sound engineer, Reg Pidsley. Their eye-witness account of the mission, complete with recordings from inside the plane, was broadcast on UK radio following their successful return. The VR experience recreates this historic event, using a mixture of real archive recordings and VR simulations, giving people a chance to experience the Berlin bombings through the eyes of those who were there. Simulations like this make historical events more accessible to those who were born long after the events took place – and bring stories to life in a uniquely human way.

Humberston Cloverfields Academy

As well as transporting us to a point in history, VR can also take students to places they may never have dreamed of visiting. Humberston Cloverfields Academy, a primary school near Grimsby, England, has built its own version of a VR classroom, with four walls made up of giant video screens. The idea is students can visit far-flung places, like the Antarctic or the Serengeti, from the comfort of their classroom – and be back in time for lunch!

One thing we can glean from this example is VR's potential to overhaul the notion of field trips and school excursions. Which brings me to . . .

Improving School Field Trips

As a parent, I know that school trips can be expensive. And I can only imagine the amount of hard work that goes into organizing the average trip. Thanks to XR technologies, particularly VR, field trips could become a lot more interesting, accessible and affordable. Imagine, for example, being able to take students on a tour of a gallery on the other side of the world, or to see polar bears in the Arctic. In theory, any kind of experience is possible with VR. And these exciting field trips could be open to a far wider range of students, not just those from affluent families.

Google Expeditions

I've mentioned a few Google examples already in this book, so you probably won't be surprised that Google is creating amazing field trip experiences for students through its Google Expeditions app. Designed for teachers to use with their classes, this epic app introduces students to a new way of learning. There are hundreds of adventures to choose from – some using VR and some using AR – spanning history, science, the arts and the natural world. In the VR experiences, students embark on an immersive simulated experience, to destinations like the Louvre or Mount Everest. And the AR version brings abstract concepts to life in the classroom. So, for example, the teacher could project a swirling tornado or beehive into the classroom, so students can get a closer look. I love how this app removes the typical barriers to field trips and opens up a whole new world of learning for young people.

SkyView

If daytime field trips are hard enough to organize, imagine how difficult it is to take students on nighttime excursions – which is the only way to go stargazing. Or is it? App developers Terminal Eleven created the SkyView AR app to help budding stargazers spot and identify celestial objects in the sky, even during the day. Students simply point a smartphone or tablet camera at the sky, and they see projections of constellations and other objects. So, they can learn to identify constellations, locate the moon, spot satellites and the International

Space Station as they zoom over, and discover distant galaxies. There's also a "time travel" feature that lets people see the sky as it would have looked in the past, or how it'll appear in the future, which is really cool. Not only do apps like this help to inspire young people, it also allows students to experience the night sky wherever they are in the world – meaning, you don't have to be atop a deserted hill at night, armed with a big telescope.

The VR Museum of Fine Art

Imagine you wanted to take art students to visit the top museums in the world and see the world's most recognizable paintings and sculptures. It'd be a pretty expensive trip, as you'd have to fly to France, Greece, the UK, the United States, Vatican City, Spain and more. But thanks to the VR Museum of Fine Art, available on Steam, you can take in all the best artworks from museums around the world, from the comfort of your home or classroom – providing you have a compatible VR headset. There's no protective glass, so you can get up close to the sculptures and paintings, which are all 1:1 scale and rendered in great detail. There's even a museum café, where you can take a virtual breather.

While experiences like this are never going to replicate the majesty of seeing a great piece of art in real life, it certainly removes some of the niggles of visiting museums, including the expense, the queues, and all those selfie sticks!

BBC Civilizations

The good old BBC is at it again with its Civilizations app, which lets users explore precious artifacts up close. Building on its epic BBC Two television series of the same name, The Beeb partnered with Nexus

Studios and around 30 museums and galleries across the UK to create the app. Through the power of AR, and using a simple smartphone or tablet screen, you can see a mummy inside a sarcophagus in the room with you, along with other artefacts from UK museums. This is great because it gives students a flavor of museum treasures without having to trek to London and other major cities around the UK.

HoloMuseum XR

With HoloMuseum, XR specialists Ximmerse created an MR educational experience for learners of all ages. Complete with a lightweight visor-style headset and handheld remote, the innovation turns any room or classroom into an interactive virtual museum. There are different educational experiences to choose from, including one with a giant T-Rex and its baby, which teaches users about the dinosaurs' diet, territory and time period. The experience made a splash at the 2020 Consumer Electronics Show (CES), with those who tested out the tech describing it as entertaining, educational and dramatic – although how accessible it is for students remains to be seen.

Unimersive

Experts in VR training and education, Unimersiv is dedicated to helping students of all ages enhance their learning through VR. They've created a number of VR educational experiences, but "A Journey into the Human Brain" – available via the Unimersiv app – is probably my favorite. It takes users inside the brain to uncover how this most complex part of the human body really works. So, you can learn about the brainstem, neurons, the cerebellum, and more – and all in an entertaining, immersive way. For me, this example shows how VR enables us to go on trips that would never have been possible before, including journeying into the human body.

Mercedes-Benz Museum

Car lovers can now take a virtual tour of the Mercedes-Benz museum in Stuttgart, Germany. Available via the museum's website, there are two main exhibitions to explore; Legend explores the history of cars, including the first models developed by Gottlieb Daimler and Carl Benz in the late 1800s, while the Collection exhibition showcases some of the most iconic Mercedes-Benz models. Users can get up close with the vehicles, including vintage cars.

Although this example is less likely to appeal to school children, it certainly shows how VR can help people of all ages indulge their passions and go on excursions around the world, without having to travel anywhere. Turn to Chapter 8 for similar examples from the world of entertainment and sport.

The Big Bang AR by CERN

Headquartered in Geneva, Switzerland, CERN is the European Organization for Nuclear Research. Here, in the world's largest particle physics lab, CERN scientists are trying to uncover the secrets of the tiniest particles in the universe, and the origin of our universe. But CERN is also working to make science more engaging and inspiring – which is the idea behind CERN's award-winning Big Bang AR app, created in partnership with Google Arts & Culture. This ambitious app uses AR to tell the 13.8-billion-year history of our universe, in only seven minutes. Oscar-winning actress Tilda Swinton provides voice-over instructions, guiding the user through the universe's evolution, from first swirly bits (that's a technical term) to the birth of stars and planets. Again, this shows how XR technologies can take us to places and times that would otherwise be impossible to visit, in order to make learning more fun and accessible. After all, learning about the Big Bang by "seeing" it is bound to be more engaging than reading about it in a textbook.

Learning By Doing: How XR Can Transform Hands-On Learning Experiences

Froggipedia

Did you ever have to dissect a frog or any other small animal in high school? I did, and I hated it. (It was worse for the poor frogs, obviously, but a little traumatic for us kids too.) Now, there's an app for that. Froggipedia, winner of the iPad App of the Year Award 2018, is an AR app that lets students study the internal organs of a frog without harming any actual frogs. You can study frog organs individually, or there's a dissection option, which lets you poke around the complex internal structure of frogs' organ systems. Plus, you get to experience the incredible transformation that frogs go through, from egg to tadpole, then from tiny froglet to a fully formed frog. There's even a fun quiz at the end. The app is available in multiple languages, including Japanese, Russian and Chinese.

This amazing app improves something that, for many, is an unpleasant learning experience, while still teaching students about the unique biology of frogs – demonstrating how technology can ease the learning progress and make something that could potentially be, well, pretty gross (there's another technical term) into a fun learning experience.

Mondly

The best way to learn a language is through immersion – spending time in the country, being among locals, immersing yourself in the language and culture, and just generally letting it seep into you. We all know this. But, in practice, most of us don't have a spare six months or a year to go off and immerse ourselves in another language. Most of us make do with some sort of DIY option – books and audio, or more recently, language-learning apps. These apps are great, up to a point, but they're not exactly immersive.

Romanian EdTech company Mondly – which provides tech-enabled language learning programs for 33 languages – set out to change that with its Mondly VR app. The idea is it's as close as you can get to immersion, without physically immersing yourself in that country. The lessons are similar to those you'd find elsewhere on Mondly, and other language-learning apps, but with the added bonus of making you feel like you're in a real situation, having a real conversation with a local – a digital woman on a train, for instance. It's obviously nowhere near as immersive as the real thing, but it certainly makes the learning experience more engaging. In theory, pretty much any kind of learning-by-doing could be improved with VR, such as learning how to service your own car or fix a leaky tap – or even taking your first few driving lessons.

Enhancing Vocational/Workplace Training and Education

Building upon this notion of learning by doing, let's explore some examples of XR-enabled vocational or workplace training and education. In the future, I believe all manner of workplace training can be enhanced through XR solutions, but it can be especially valuable in simulating dangerous situations or scenarios that are difficult to simulate in real life.

FLAIM Systems

Could VR train the next generation of firefighters? That's certainly the idea behind Australian company FLAIM Systems' technology. Australia has been devastated by wildfires in recent years, as have parts of America. Now, some fire departments in Australia and the United States (alongside the UK, Netherlands and other countries around the world) are using FLAIM's VR technology to train their firefighters, immersing them in virtual scenarios that would just be too dangerous or difficult to recreate in the real world. These scenarios include wildfires, as well as house fires and airplane fires, and the VR tech realistically renders all the fire, smoke, water and fire-extinguishing foam – as well as the heat, thanks to a special heat suit that can heat a firefighter up to around 100 degrees Celsius (212 Fahrenheit), depending on their proximity to the virtual fire. The system even replicates the powerful force firefighters feel from the water hose. The technology is so impressive, the Australian tech industry named FLAIM Systems startup of the year in 2019. As well as making training safer and more immersive, FLAIM's technology also reduces the environmental impact of training firefighters, since there's no actual smoke or water being used.

This is an incredible example of VR being used to improve training and realistically train people in a wide variety of scenarios that they might otherwise not be able to experience in a traditional training program.

University of Exeter and Cineon Training

When it comes to high-risk jobs, nuclear engineering must be right up there. Training in a field like this can be extremely difficult and expensive – and there's only so many training scenarios you can run in the real world without prompting nuclear disaster. Which is why researchers at the University of Exeter, England have teamed up with immersive learning specialists Cineon Training to create VR training programs that will soon be used to teach nuclear engineers. Not only can engineers train in a variety of settings, it also means engineers can be trained before nuclear facilities have even been built.

BP

British multinational oil and gas company BP is one of the world's leading energy providers. It's also a company that embraces technology – especially big data and artificial intelligence – to improve operations.

So it's no surprise to me that BP is also investing in VR-enabled training for employees working in dangerous situations. To train employees in startup and emergency exit procedures at BP's oil refinery in Hull, England, the company partnered with Igloo Vision – known for creating immersive shared VR spaces. When you work in an oil refinery, mistakes can be fatal, but the virtual training allowed employees to learn from their mistakes safely. How they pulled off this training is particularly interesting; rather than trainees each wearing their own VR headset, Igloo Vision built a six-meter igloo at the Hull refinery. Inside the igloo, employees can experience an extremely detailed replica of the plant and practice critical safety tasks, all in a safe, virtual setting. This is great because it provides the opportunity to assess whole shift teams at a time, rather than immerse each individual in their own simulation – which could be the future for critical team-based training exercises and assessments.

New Jersey Police

Law enforcement officers in Camden County Police Department, New Jersey are using a VR simulator to train for more than 230 real-world scenarios they may encounter – covering everything from a routine traffic stop or domestic violence callout to mass shooters. In the 360-degree simulations, officers learn how to de-escalate dangerous situations wherever possible, and when to use force. The technology is created by VirTra, providers of simulation training in the judgmental use of force, and is used by a range of military and law enforcement agencies. For more examples like this, all from the worlds of law enforcement and the military, turn to Chapter 12.

STRIVR

It's not just high-risk jobs that benefit from VR training. Athletes are also beginning to make greater use of the technology. It makes perfect sense, when you think that much of any sports-training regimen relies on repetition, whether it's on a football field, tennis court or wherever. But getting those reps in can be a challenge – particularly if the weather is bad, or you're traveling, or even injured. Immersive learning specialists STRIVR promise to change all that with their immersive learning technology for sports, which was developed in collaboration with the Stanford University Football team. The technology can be applied to a range of sports, but in the case of football, it allows athletes to be transported to the football field, from their own home (or wherever they are). Using the virtual simulations, players can learn and practice customized plays (of which there could be hundreds of different variations to memorize). STRIVR says its technology is now being used by many NFL and college football teams to accelerate player development. You can find more sport and entertainment examples in Chapter 8.

LAP Mentor and immersive medical training

VR and AR are also finding a lot of application in medical training, particularly surgery. This reduces the need to train on real-life patients (with all the risk that brings), while still giving trainee surgeons a relatively realistic simulation of operating rooms and procedures. LAP Mentor, designed to improve laparoscopic surgical skills, is one such example. The trainee puts on a VR headset and is then transported to a virtual operating theater, complete with all the equipment and even the noises. Using hand controllers, the trainee can learn basic laparoscopic skills and complete laparoscopic operations. There's also a built-in assessment feature that gives the user feedback.

When combined with the supervision and feedback of a human instructor, it's easy to see how tools like this aid the transference of skills and help to build confidence for those taking part.

Children's Hospital in Los Angeles (CHLA)

In a similar example, CHLA partnered with AI and VR specialists AiSolve, Oculus and BioflightVR to create a virtual training solution for pediatric surgeons. The simulation is reportedly so detailed, it includes simulated versions of the hospital's real-life nurses to ensure trainees' experience of the virtual operating theater would match that in the real one. I love this idea of digitally recreating actual personnel in order to make a simulation more immersive and realistic.

The Body VR: Journey Inside a Cell

Finally, here's another example of a VR experience for medical students – one that lets you take a trip inside the human body and travel through the blood stream. This award-winning educational app allows you to explore the billions of living cells in the human body and learn how blood cells spread oxygen through the body and how the body reacts to deadly viruses – all in a visually impressive, engaging way. It's designed for medical training, with the goal of improving healthcare outcomes, but I can see how it'd be fascinating for anyone interested in how the body works.

Lessons We Can Learn from Training and Education

After exploring all these examples, it's clear that educational XR experiences present a huge business opportunity. From elementary school right up to university level, lifelong learning and corporate training, XR can make education and training more effective and relevant in today's tech-driven world. This will mean different things to different organizations. Your business may be excited about the opportunities to create and market your own educational offerings, or you may feel now is the time to overhaul your in-house training to become more immersive. Either way, the examples in this chapter highlight five common factors that make for a powerful VR or AR learning experience.

- It must be easy to use. Regardless of the age group, users must be able to master the technology intuitively. The goal is to ease the learning journey, not make it harder and longer.

- It must be as immersive as possible. For me, the most powerful examples in this chapter are those that really bring subjects to life. Ideally, your experience will give the user the feeling of being *in* the experience.

- It must provide some measure of progress. From teachers monitoring how well students are progressing, to corporate trainers assessing the competence of trainees, you must be able to measure the impact of your experience. What does success look like for your users? How will you monitor and track this?

- It should tell a good story. As well as being immersive, the best learning experiences engage users with an element of storytelling.

- It will ideally adapt to users' abilities. Some of the more advanced XR educational experiences track how the user is learning and then tailor the content to the user's abilities – for example, by slowing down and focusing on various aspects when a user is struggling.

I also firmly believe that such experiences should be as accessible as possible, so that they can be available to more people. The danger with XR technology, as I mentioned in Chapter 3, is that it could widen social divides, which is not something we want to exacerbate in the world of education (and I mean education at all stages of life). Therefore, if you're creating an XR educational experience, I urge you to consider affordability and accessibility, and create experiences that work on devices people already have or that are affordable to buy (so, tablets, phones and low-cost VR hardware, like Google Cardboard, which you can get for as little as $10). Insisting on expensive hardware creates a significant barrier to entry.

Key Takeaways

In this chapter, we've learned:

- The EdTech sector is growing rapidly, meaning there are mouth-watering business opportunities to provide immersive, engaging educational experiences for learners at all stages of life, from young school-age children to college students to lifelong learners.

- VR is being used in formal education settings to provide more immersive learning experiences, offer more exciting (and easier and cheaper) school field trips, and enhance learning-by-doing exercises.

- Meanwhile, AR is helping to bring concepts to life for students, allowing them to visualize concepts and interact with learning materials in new ways.

- Evidence suggests that students who learn through such immersive tools outperform those who don't – thus, XR technologies make it easier for students to comprehend and retain knowledge. Plus, the students have more fun in the process!

- VR and AR are also beginning to transform the world of corporate training, particularly when it comes to simulating training scenarios that would be too difficult, dangerous or expensive to simulate in the real world.

I particularly love the examples in this chapter that enhance training for high-risk and difficult jobs, including medicine. In the next chapter, I dig deeper into the world of healthcare to see how XR is being used to improve patient care and enhance the work of health-care professionals.

Endnotes

i. Smart Education and Learning Market Size, Share & Trends Analysis Report; Research and Markets; https://www.researchandmarkets.com/reports/4621713/smart-education-and-learning-market-size-share?utm_source=dynamic&utm_medium=GNOM&utm_code=gqnhtl&utm_campaign=1397658+-+Global+Smart+Education+and+Learning+Market+(2020+to+2027)+-+by+Age%2c+Component%2c+Learning+Mode%2c+End-user%2c+Region+and+-Segment+Forecasts&utm_exec=jamu273gnomd

ii. Active Learning to improve long-term knowledge retention; Academia; https://www.academia.edu/1969321/Active_Learning_to_improve_long_term_knowledge_retention

iii. Glossophobia or the Fear of Public Speaking; Very Well Mind; https://www.verywellmind.com/glossophobia-2671860

iv. A Structural Equation Modeling Investigation of the Emotional Value of Immersive Virtual Reality in Education; ResearchGate; https://www.researchgate.net/publication/322887672_A_Structural_Equation_Modeling_Investigation_of_the_Emotional_Value_of_Immersive_Virtual_Reality_in_Education

v. Virtual Learning Simulations in High School; Frontiers in Psychology; https://www.ncbi.nlm.nih.gov/pmc/articles/PMC5447738/

vi. Cristal Steel using MEL Chemistry VR in class; LendED; https://www.lended.org.uk/case-study/cristal-steele-using-mel-chemistry-vr-in-class/

vii. WCU-LA Anatomy Students See Full Letter Grade Improvement Using HoloLens; West Coast University; https://westcoastuniversity.edu/pulse/health-e-news/wcu-partners-with-microsoft-to-integrate-augmented-reality-into-student-learning-experience.html

7

HEALTHCARE

In the last chapter I talked about the incredible opportunity we have to transform education through technology. The same is certainly true of healthcare. In fact, as the use cases in this chapter show, there's huge potential for XR technologies – namely, VR and AR – to improve all aspects of healthcare, from self-care and well-being to diagnosis and treatment, and even surgical procedures. It's early days, and adoption of XR in healthcare may be more gradual than in other sectors described in this book – after all, it's important that any new healthcare solution is implemented in an evidence-based way, and getting official approval for new medical technologies can, rightly, take some time – but, believe me, the transformation is already under way.

As an example, AR can be used to visualize medical data – by overlaying anatomical data (for example, a map of veins) onto the patient in real life, to help clinicians carry out procedures in a faster, more accurate way. Or navigation data in non-invasive surgeries can be projected onto a transparent head-up display that the surgeon is wearing, meaning they don't need to look away from the patient to view on-screen data or images – the information can be projected in their line of sight, or potentially even onto the patient. As we've already seen in this book, AR excels at bringing information to life in new, richer ways – why shouldn't that include medical information?

VR, meanwhile, is demonstrating its therapeutic chops by, among other things, immersing patients in relaxing simulated environments, which can help to calm them before (or even during) treatment, reduce pain and generally improve the experience of being in a hospital or clinical setting. And that goes for patients of all ages – from children getting their latest vaccine to women in labor to elderly patients suffering from dementia.

VR is also being used to enhance medical training (as mentioned in the previous chapter). Nowadays, we have VR-based training solutions that are more *Top Gun* than *Grey's Anatomy!* Take FundamentalVR as an example. Named by Time Magazine as one of the best inventions of 2018, and accredited by the Royal College of Surgeons of England, FundamentalVR's technology acts like a flight simulator for surgeons, allowing them to practice their surgical techniques in a safe, controlled environment.

If this sounds a bit futuristic or sci-fi, it's not. The use cases in this chapter are all real-life examples from current clinical and research settings. It's with good reason, then, that AR and VR are frequently described as breakthrough technologies in healthcare.

And, boy, do we need breakthroughs in healthcare. The population on our planet is growing and people are, generally speaking, living longer and longer. Healthcare services around the world are under increasing pressure, waiting times can be long, access to certain services (such as mental health services) can be limited, and, depending on where you are in the world, medical treatment can be hugely expensive. VR and AR can not only improve patient outcomes by making treatment more effective, they can vastly improve accessibility for healthcare services and make healthcare more affordable. While some of the cases in this chapter involve enormously expensive proprietary technology, many make use of readily available, off-the-shelf technology, such as inexpensive VR headsets. With technology like this, patients can even continue treatment at home, between therapy sessions. Which is particularly pertinent at the time of writing this book.

With the advent of stay-at-home orders and social distancing rules, the coronavirus pandemic certainly accelerated the need to successfully treat and monitor patients remotely. If you're around my age, you may remember doctors making house calls when you were a child – which seems extremely quaint and old-fashioned these days. Nowadays, house calls are virtually non-existent, and most people are expected to attend the doctor's surgery. Although, of course, not at the time of coronavirus. During the pandemic, video and telephone consultations became the norm for many, and it's reasonable to expect an acceleration of remote patient interactions after the virus. VR and AR can help to smooth such changes, allowing remote VR-based therapy sessions or even through the creation of VR clinics, where patients and doctors can interact in the same virtual space.

Over time, our encounters with healthcare professionals will increasingly be shaped by XR, especially as the technology becomes more and more affordable. But for now, let's delve into some of the best current use cases from healthcare.

Using XR to Enhance Your Well-Being

In our ever-connected society, there's growing concern about the impact technology has on our ability to relax and "switch off." Yet, perhaps ironically, technology may prove part of the solution. That's certainly the idea behind the growing range of XR-enabled solutions in the wellness and well-being space. Let's take a look at some of my favorite examples.

VR-enhanced relaxation – the Dream Machine

Mindfulness may be good for stress and anxiety, but training your brain to be *in the moment* – especially when you're feeling stressed – is easier said than done. So Lebanese-born tech expert Dr. Jamil El-Imad, founder of NeuroPro, used his knowledge of cloud computing, VR and neuroscience to create an experience that helps. Called the Dream Machine, it's a combination of a VR headset and electro-encephalogram (EEG) headset, which monitors electrical activity in the brain – meaning the system can measure the subject's response to relaxing images and help them maintain relaxation.

After putting on both headsets – which isn't as cumbersome and bulky as it sounds (both use off-the-shelf technology, rather than being specially designed for the Dream Machine) – the user hears relaxing music and sees waves lapping on a beach on a virtual island, complete with giant Easter Island statues among the palm trees. White feathers float on the breeze, and the user's breathing keeps them afloat. The idea is the user focuses on the moment at hand, keeping the feathers floating and looking at the statues' faces. When the user's mind starts to wander and they lose relaxation, a white fog appears and blocks their view, which signals that they're losing focus. The idea is this trains the brain to be more present in the moment. And afterwards, the user is given a score that represents their concentration level – giving them a baseline to measure improvement. In the future, systems like this could prove useful in a number of scenarios, such as reducing anxiety and stress in workplaces or just generally helping people improve their mindfulness practice.

Guided meditations – with added VR

Taking mindfulness and relaxation to the next level, there is a plethora of meditation apps and classes out there to help you meditate and de-stress, often using relaxing sounds and guided instructions.

Some of these solutions are now incorporating VR to make guided meditations more immersive. If you've ever struggled to tune out the real world while meditating, you can probably imagine the benefits of this. With a VR headset, you can literally block out what's going on around you and immerse yourself in a digital meditation space, complete with relaxing visuals.

Guided Meditation VR is one such app. It comes with more than 100 ready-made meditations, or you can choose to customize your own experience. So, you could be meditating on a beach, or in a secluded forest, or even on top of a mountain – meanwhile, gentle audio instructions guide you through various breathing exercises.

The Provata VR guided meditation and mindfulness app does a very similar thing, but with the added bonus of tracking your health indicators (such as your heart rate) to monitor your progress and provide smart feedback. No yoga instructor in the real world would be able to track the heart rates of all participants in their class.

Or, if you prefer a more communal meditation experience, EvolVR's weekly virtual meditation sessions might be for you. These free group sessions take place on AltSpace VR, Microsoft's social VR platform, and are led by EvolVR founder and ordained Unitarian Universalist minister Jeremy Nickel and his team. Participants get to introduce themselves at the start of each session, building more of a sense of community, then there's a guided meditation for everyone to follow. Although each participant is in their different geographic locations, you feel like you're together in one place – and this supportive environment may help those who struggle with meditation or who don't enjoy the solitary nature of meditating alone.

I love how these very different approaches to meditation are being enhanced through VR, demonstrating how a range of wellness techniques and practices could potentially incorporate XR technologies in future.

XR and yoga

Learning yoga at home is tricky. Sure, there are some brilliant You-Tube tutorials and yoga apps that you can use, but you don't get the same benefits as you would in a class, with a teacher there to guide you and make sure you're properly positioned. Trouble is, many people aren't able to attend or afford such classes, so they make do with the yoga-at-home alternatives. Now, yoga apps are beginning to incorporate XR technology to help people get more out of their at-home practice.

One brilliant example comes from Brooke Schuler's Introduction to Yoga in Augmented Reality kit. As well as a fully illustrated book, you get an AR-enabled app with a virtual instructor projected into your room to demonstrate poses. Not only does this make the tutorial more interactive and interesting, it makes it much easier to understand and emulate each position. There's even been some scientific study on incorporating AR into yoga, which found that the application of AR can help participants enhance their perception of the yoga experience.[i]

Taking this to the next level, we may see yoga classes go virtual, just like the communal guided meditation I just mentioned. Here, you would join a yoga class virtually, potentially with people from all over the world, and, thanks to your virtual avatar, the instructor would be able to see how you're doing and offer feedback to improve your downward dog.

VR fitness coaching and workouts

Could your next personal trainer be a VR app? It could well be, if Supernatural is anything to go by. Designed to improve home workouts, Supernatural's VR workout experiences combine music, motivating virtual coaches and beautiful virtual destinations to create workouts that you actually look forward to. The company releases a new VR workout every day, so you never get bored, and you can exercise in some of the most breathtakingly beautiful settings in the world, without having to leave your home.

There are also VR games that double as exercise, like ICAROS VR. It's essentially a VR gaming platform, but one that gets you off the couch and moving around, doing different virtual activities like swimming, flying or racing. Your real-life body controls the movements in the game, which all helps to improve core muscles and balance. (For more sport-related XR examples, turn to Chapter 8.)

As well as making exercise more enjoyable, there's some evidence that VR also helps people exercise better. Research by the University of Kent, England, shows that VR can improve exercise performance, reduce pain and increase the amount of time people can sustain an activity.[ii]

Never mind making exercise at home more fun; VR workouts could even replace the sterile gym environment. If I ran a gym, I'd be seriously eyeing VR technology and wondering how I could make the gym experience more immersive and fun – how I could use VR to help members push themselves for longer, experience less discomfort, and achieve their fitness goals in an easier, more enjoyable way.

Use of XR in Diagnosing Health Issues

XR is also making huge strides in the diagnosis of illness and health issues, from mental health conditions to physical trauma, injury and disease. There are many advantages to using VR-based or AR-based diagnostic tools, most notably that off-the-shelf technology is readily available and is significantly cheaper than neuroimaging, meaning VR and AR can be deployed in a wider range of clinical settings. Let's take a look at some real-world examples that demonstrate VR and AR's potential as a diagnostic tool.

Medical imaging and analysis

When you think about it, medical imaging is a form of augmented reality, since it helps to visualize the body's internal structures in real time. But now, AR can be used in a more literal sense to improve medical imaging, by creating more integrated and interactive visual displays of medical data for analysis – for example, by projecting internal structures such as vessels or tumors onto a display, headset, or even onto the patient themselves (useful in the context of prepping patients for surgery). VR, too, can provide exciting new ways to display medical images, by giving the healthcare professional (or patient) the feeling of being "inside" the image.

Developments like this are important because the vast majority of medical data is in the form of images, like scans and x-rays (some estimates suggest up to 80 or 90 percent of medical data is image-based). If we can improve the way we view and analyze visual medical data, then we should be able to improve patient outcomes.

One exciting example comes from DICOM VR, which was founded by medical physicist Chris Williams and radiation oncologist Kos Kovtun in response to the increasing complexity of medical imaging. As they put it, medical imaging is currently restricted to displaying 3D information on a 2D screen; yet, using VR, Williams and Kovtun are developing a system that allows clinicians to intuitively view and manipulate images in a 3D VR environment. The idea is this will improve the speed and precision of image-guided cancer diagnosis and treatment.

Diagnosing visual impairments

Another great diagnostic example comes from SyncThink, a company founded in 2009 to develop mobile eye-tracking technology. The firm's EYE-SYNC tool uses VR goggles with eye-tracking technology to measure visual impairment and help clinicians diagnose whether someone has a concussion. The system has been approved by the US Food and Drug Administration since 2016.

VR has also been used to spot early signs of glaucoma. A team at the Krembil Research Institute, in Canada, used the popular Oculus Rift headset to detect "vection" in participants (vection being the sensation we feel when our field of vision is moving and it feels like we're moving too, even though we're not). In patients with early or mild glaucoma, this vection sensation is impaired or absent.

Another team, this time from Cardiff University in Wales, has been investigating the use of VR to diagnose and treat visual vertigo (or persistent postural perception dizziness, to give it its full name).[iii]

All these examples hint at the potentially huge range of physical conditions that could be diagnosed using VR.

Aiding psychiatric, neurological and mental health diagnosis

In Chapter 6, I talked about how VR can help people overcome their social anxiety around public speaking; yet, VR can also be used as a diagnostic tool for social anxiety, by analyzing participants' gaze behavior (because findings show that socially anxious people tend to focus on the facial region for shorter periods of time).[iv]

However, when it comes to diagnosing psychiatric conditions, like schizophrenia, diagnosis is notoriously difficult – which is partly why, according to the World Health Organization, between 35 and 85 percent of mental health conditions go undetected and undiagnosed.[v] In one example, a team from the University of Exeter, in England, successfully used a VR-based "mirror game" – in which participants had to copy the gestures and facial expressions of a virtual avatar – to aid the early diagnosis of schizophrenia.[vi]

VR has also been trialed as a way to detect early signs of Alzheimer's disease. In one study, led by Dr Dennis Chan of the University of Cambridge, researchers used a VR headset to test participants' spatial navigation and memory. (The brain contains a "mental satnav," if you will, which knows where we are, where we've been and how to find our way. This is one of the first regions to be affected by Alzheimer's disease, which is why getting lost is often an early sign of the disease.) Results from the study showed the VR navigation test was more accurate at spotting mild or early Alzheimer's-related impairments than traditional "gold-standard" diagnostic tests,[vii] which is incredible.

For me, the beauty of VR as a diagnostic tool is it can create realistic simulations of real-world situations or scenarios that may provoke symptoms, that would otherwise be impossible or very difficult to simulate in a clinical setting – and, importantly, VR can do this in a controlled, consistent way, which could help clinicians make more objective diagnoses. It's also cheaper than neuro-imaging techniques and can be deployed in a wider range of settings.

XR in Treatment and Therapy

Moving from diagnosis to treatment, let's see how VR in particular is being used to improve therapeutic treatments – across both physical and mental health – and create better outcomes for patients.

VR Vaccine – helping children overcome their fear of treatment

Children often fear medical treatment, including anything from trips to the dentist and routine vaccinations to more serious surgeries.

Now, VR is being used to help young patients overcome their fears, by distracting them with immersive stories.

In one example, a team in Brazil used VR to help children beat their fear of vaccinations. The project, called VR Vaccine, involved children watching (via a VR headset) an animated adventure story, while a nurse (who can see the story unfolding on a separate screen) synchronizes the action of cleansing the skin and administering the injection with the story. The team's research showed that most children feared the needle itself, rather than the pain they might feel, so the VR approach was devised to essentially block out and distract from the needle. It also helped the children relax, making it easier for the nurse to give them the shot. The project was the brainchild of Brazilian pharmacy chain Hermes Pardini, in collaboration with local design studios VZLAB and Lobo – and it proved so successful that Hermes Pardini has since installed VR headsets in all of its pharmacies to help with its vaccine campaigns.[viii]

For me, this example clearly shows VR's ability to relax patients and help them cope with anxieties around treatment – something that could prove useful for patients of any age, not just children. (Indeed, there are similar examples related to surgical procedures, as we'll see later in the chapter.)

Virtual reality exposure therapy (VRET) for anxiety and PTSD

VR is beginning to establish itself as a useful tool in mental health treatment – and the American Psychological Association has reported that VR is "particularly well suited to exposure therapy."[ix] This brings us to virtual reality exposure therapy (VRET) which is, at the time of writing, mostly being used to treat post-traumatic stress disorder (PTSD) and anxiety disorders. As the name suggests, VRET involves using VR to expose patients to scenarios that may trigger their anxiety or PTSD symptoms, all within a controlled, safe environment. The goal is to condition the patient to confront triggers, process the emotions that arise as a result, and engage more deeply with treatment. For mental health providers, the obvious advantage is that is allows clinicians to simulate scenarios that would be otherwise challenging to recreate, and to control every element of the patient's exposure (all in a cost-effective way). Patients benefit by also having a greater sense of control than they would if they were attempting to confront triggers in the real world – plus, treatment can be continued at home, at their own pace.

Evidence is emerging to support this use of VR in exposure therapy. Researchers from Emory University in Atlanta used VRET to treat veterans suffering from PTSD as a result of military sexual trauma. (Participants were shown simulated VR clips of military bases.) And according to the findings, not only is VRET safe to use in these circumstances, it helps to reduce depression and PTSD symptoms.[x] VRET is also being explored to treat substance abuse problems.[xi] And in the UK, Norfolk and Suffolk NHS Foundation Trust, part of the National Health Service (NHS), has adopted VRET for treating a range of phobias, including fear of flying, fear of heights and fear of spiders.[xii]

This isn't to say VRET will replace conventional therapeutic solutions, but it does demonstrate how VR can provide an additional boost to more traditional approaches.

VR-enhanced cognitive behavioral therapy

Of course, exposure therapy isn't the only therapeutic treatment for anxiety disorders, PTSD and other mental health conditions. Cognitive behavioral therapy (CBT) is another widely used, highly effective approach.

Oxford VR, a company that spun out of Oxford University, are leaders when it comes to combining clinical psychological science and VR. In 2020, the UK's NHS announced that it was offering Oxford VR's social engagement treatment program to patients with social anxiety.[xiii] The program takes a CBT-based approach, but applies it in an immersive virtual setting – meaning patients put on a VR headset and are then guided through a series of tasks and settings that might trigger anxiety and social avoidance, such as being on a busy street or using public transport. Gradually, patients learn to face problematic scenarios and build their confidence by trying out tasks without any risk of harm. Crucially, the program can be delivered by any trained member of staff, not necessarily a qualified clinician – potentially making treatment much more accessible when waiting times for mental health appointments can be long. And by Oxford VR's own claims, their technology is not only highly effective – it delivers fast results, too. As an example, Oxford VR's fear-of-heights simulator reduced fear by an average of 68 percent after only *two hours* of treatment.[xiv]

In 2020, Oxford VR was awarded £10 million in venture capital funding, demonstrating that confidence in VR-led therapies is high. In other words, it's a good time to be in the medical technology business.

Treating psychosis with VR

The gameChange project – a collaboration between Oxford University's Department of Psychiatry and the NHS, along with multiple other organizations – is exploring the use of immersive VR to help patients with psychosis. The idea is patients put on a VR headset, and then a virtual coach takes them through simulations of scenarios they find troubling, helping the patients practice techniques to overcome difficulties. Importantly, the program is automated, meaning it's led by the virtual coach, allowing for potentially very widespread use in the NHS. This is really exciting, because it could dramatically increase the number of people who can access therapy, and reduce the amount of time they have to wait to access that therapy.

Helping patients with dementia

Worldwide, around 50 million people suffer from dementia, and millions of new cases are diagnosed every year. Earlier in this chapter, I mentioned how VR has been trialed in diagnosing Alzheimer's disease, which is the most common form of dementia. Yet VR can also help to improve quality of life for dementia patients. As dementia is an incurable disease, most available medications focus only on improving symptoms temporarily, so anything that can help to improve patients' quality of life is welcome.

In one example, patients at Marston Court care home in Oxford, England, were given VR headsets to take them back in time to revisit favorite hobbies, places and formative times in their lives. One patient, for example, visited France, where she had run a bed and breakfast with her family. Another was transported back to the rock and roll era. Patients reported feeling joy and comfort from reliving their memories and times gone by.[xv]

In another example, health technology startup VR Revival has created a VR app that aims to improve quality of life for dementia patients in Africa and reduce the stigma surrounding dementia in Africa. Like the previous example, this app provides immersive VR experiences that are designed to uplift patients.

Helping children with autism

Young people with autism often need a lot of additional support at school and in everyday life – some of which could come in the form of VR tools.

The Mendip School in Prestleigh, England, collaborated with the University of the West of England and VR specialists Go Virtually in a VR research project designed to help students with autism learn new social skills and build their confidence. Named Virtual Reality Technology, and used by autism groups, the project worked with students aged 6 to 16 to trial different VR technologies. Students reported that they most liked to use the technology for relaxing and meditation, learning about places they hadn't been before and reducing the anxiety associated with going somewhere new.[xvi] One thing that I really liked about this project is that it determined that low-cost tech like Google Cardboard, coupled with a smartphone, was a good first step into using VR with autistic groups – meaning expensive high-tech equipment isn't needed.

VR-enhanced rehabilitation

Depending on the injury, physical rehabilitation can be a long, slow, frustrating and painful process. The Defence Medical Rehabilitation Centre in Loughborough, England – which provides advanced specialist care for injured military personnel – is using VR technology to aid that process.

The center's Computer Assisted Rehabilitation Environment (CAREN) machine is essentially a big treadmill surrounded by screens – a bit like a flight simulator, but one that surrounds the patients with VR images as they walk on the treadmill. These screens can re-create different real-world situations, such as walking on tricky cobbles. Meanwhile, equipment and sensors monitor the patient's muscle activity to see how muscles are being used. Patients can also watch their muscles move, understand which muscles are working well and which ones aren't, and see how their muscles are improving with physical therapy.

In the future, VR simulators like this could be rolled out in various different clinical settings to help patients' rehabilitation.

Managing pain

Pain management is a critical part of treating physical injury and illness, and this is another area that can be enhanced through VR. Clinical trials have shown that VR helps patients cope better with pain and reduces the use of opioids – including one study where patients treated with VR reported a three-point reduction in pain definition, on a scale of 1–10.[xvii]

Hoag Hospital in California is one of the United States' first hospitals to use VR pain treatment (outside of clinical trial settings, that is) to help patients feel better. The coronavirus pandemic hit just as the hospital was beginning to implement the VR technology, which meant even non-Covid patients were sitting in isolation, without visitors. Using VR, these patients were able to immerse themselves in beautiful destinations or exciting experiences, like swimming with dolphins – the idea being this would relax patients and distract them from the pain. Did it work? After a six-week period, treating around 200 patients with VR, the team reported amazing results, with some patients saying the VR therapy was better than morphine.[xviii] As well as asking patients to rate their pain before and after the VR sessions – which lasted from 15 to 30 minutes – the team also took MRIs before and after some of the sessions. The MRI data showed that the brain's response to pain decreased while being treated with VR.

I find this use case really encouraging, not just because it shows how VR helps people suffering from chronic pain (just think of how many people living with chronic pain could benefit from low-cost VR therapy, without having to leave the house), but also that it could help to reduce opioid dependence in future.

Improving Surgeries Through XR

Now let's move into the operating theater, and explore the use of both AR and VR to improve surgical procedures and patient outcomes. These technologies are beginning to be deployed in a number of ways, from helping patients relax during procedures to enhancing surgical training and even using XR during procedures to help surgeons visualize the procedure and better monitor patients' vital signs during surgery (remember, AR overlays digital information onto the real world, meaning it can be used to project visual data, such as nerves and vessels, onto the patient's body).

Studies suggest that surgeons are increasingly interested in using XR technologies, particularly AR, to improve the safety and efficacy of surgical procedures, and that AR systems match the performance of traditional techniques.[xix] So, with that in mind, let's look at some inspiring use cases of XR in surgical procedures.

Reducing patients' stress

From some of the examples we've already seen in this chapter, it's clear that VR in particular is very effective at distracting and relaxing people. For patients under regional anesthetic (i.e., they aren't "put under" for the procedure), VR could help them stay calm and relaxed during surgery.

That was the idea behind a pilot study at St. George's Hospital in London. Here, patients undergoing procedures with regional anesthetic were given the option of using a VR headset before and during their operation. Those patients who used the technology were immersed in calming virtual landscapes, which proved incredibly effective. A staggering 100 percent of participants said wearing the headset improved their overall hospital experience, 94 percent said they felt more relaxed and 80 percent reported feeling less pain.[xx] Patients reported feeling so immersed in the experience, they weren't even aware of being in the operating theater.

VR has also been used to help relax women in labor and help them cope with the pain of childbirth. University Hospital of Wales in Cardiff has been trialing the use of VR headsets as an alternative form of pain management during labor.

These use cases show how VR could be used to help patients during a range of medical interventions – essentially, almost any kind of procedure where the patient remains awake, or just generally to make hospital stays less stressful for patients. (I'm also thinking it could make visits to the dentist far more pleasant!) As such, VR could potentially help to reduce the use of sedatives, or the need to put patients under a general anesthetic, which, after all, requires a longer recovery time.

Vein visualization with AccuVein

Founded in 2006, AccuVein is the global leader in vein visualization – which is where a map of veins is overlayed on the surface of the patient's skin to help health professionals find veins more easily (for starting IVs and drawing blood). Since the technology was first developed, adoption has been steadily growing, and the company says it has now helped treat more than 10 million patients. The technology is primarily used to help clinicians find veins that otherwise couldn't be seen or felt, and evidence shows that vein visualization dramatically improves clinicians' ability to find veins on the first attempt – by as much as 98 percent in pediatric cases and 96 percent with adult patients.[xxi] The technology has also been adopted to help cosmetic practitioners avoid veins when administering Botox and filler injections.

Medivis and presurgical data

Medivis is just one of many companies looking to incorporate AR and VR into operating procedures. But what's interesting about Medivis is they combine AR with AI (artificial intelligence) to provide more intelligent information and better insights, in 3D form. The idea is the AR technology helps surgeons accurately plan and perform surgery with greater precision, because they aren't having to rely on 2D imaging technology to visualize presurgical patient data, while the AI provides key insights that lead to better decision-making during surgery. This combination of technologies is particularly interesting to me, and could hint at where cutting-edge XR technology is headed in future.

Surgical Theater's Precision VR

Surgical Theater was founded in 2020 by two Israeli Air Force officers and flight simulator experts who had an interesting idea – what if surgeons could train in simulators, like fighter pilots do? Today, the company's Precision VR "surgical rehearsal platform," which is designed specifically for neurosurgical procedures, has been used by the Mayo Clinic, UCLA School of Medicine, Stanford School of Medicine, and more. The goal is to provide better preoperative planning for surgeons, by turning conventional 2D medical data into patient-specific VR simulations of procedures. But what makes Precision VR unique is it also allows patients to view a simulation of their surgery before it takes place.

I love how this technology is designed with both surgeons and patients in mind, providing a common language for clinicians and their patients (and patients' families), so that everyone has a much better understanding of the procedure that's about to take place.

SentiAR – holographic visualizations during surgery

Much as we might use a satnav system to navigate a new city, surgeons use technology to guide their movements and view the patient's anatomy during surgery. One example of this comes from SentiAR, which uses AR to create a holographic 3D visualization of a patient's anatomy that floats over the patient during surgery. The technology – which is controlled by a holographic headset, meaning it's "hands-free" for the clinician – converts 2D scan data and real-time mapping data into a hologram in the clinician's field of view. Designed especially for cardiac models, the technology aims to make cardiac procedures significantly faster and more accurate.

AR visualizations in spinal treatments

In a study of 42 spinal procedures, AR was used to provide image data before and during procedures. The image data were used to automatically segment the vertebrae and assign unique colors to each one. This data was then displayed on heads-up displays to help surgeons with anything from realigning the spine to removing tumors. It was also used to help in the training of residents. All tests were successful, according to the study.[xxii]

And, in 2020, Augmedics' AR xvision Spine System – which has been cleared for use by the FDA – was successfully used for the first time in spinal fusion surgery in the United States.[xxiii] The guidance system, which was used by surgeons from Johns Hopkins University, allows surgeons to visualize the spinal anatomy of the patient in 3D, as if they were wearing x-ray glasses. (The system consists of a transparent near-eye-display headset, so it doesn't interfere with the surgeon's vision.) This meant surgeons could accurately navigate instruments and implants while continuing to look at the patient, instead of having to look away at screens.

This is a useful indicator of how AR could be used to improve a range of procedures in future, by giving surgeons the information they need directly in their field of vision. Indeed, Augmedics plans to move into uses beyond spinal surgeries in the future, so watch this space.

Combing VR with surgical robotics

One particularly cutting-edge area of VR in surgery involves combining the technology with surgical robotics. That's the idea behind Vicarious Surgical, which was founded in 2014 to create human-like surgical robots that can perform minimally invasive surgery. By combining robotics and VR, Vicarious Surgical claims its technology transports surgeons inside the patient (virtually speaking). Theirs was the first surgical robot to receive FDA Breakthrough Device designation, which recognizes breakthrough technologies that could provide more effective treatment, and the company counts Bill Gates among its investors.

I'm always excited to see different technologies converge in interesting use cases, so the combination of VR and robotic surgeries should be an interesting one to watch in future.

Lessons We Can Learn from Healthcare

It's clear from this chapter that confidence in healthcare-related XR technology is high. Clinicians appear to be welcoming the adoption of VR and AR in clinical settings, and a wide range of evidence is emerging to support such uses. This all means there are exciting business opportunities for companies in the medical technology field, or those looking to create healthcare- or wellness-related XR tools. But what lessons can we learn from current applications? For me, the healthcare sector teaches us that:

- Given the virtual nature of VR experiences, you'd think that VR is suited solely to mental health and wellness. But I am encouraged to see how VR is also being successfully applied to physical treatments with great success (in pain management, for example). This is an important reminder of how the virtual world and the real world are becoming increasingly intertwined, and the distinction between the two is becoming blurrier.

- The combination of XR technologies with other technologies, such as AI or robotics, is particularly powerful and exciting. Consider whether your own XR solutions could intersect with other cutting-edge technologies to deliver better insights and improve user outcomes.

- There's no denying that some of the solutions in this chapter – like the VR surgical simulators I mentioned – are hugely expensive. But that's not always the case. Many of the examples, trials and projects quoted used inexpensive, off-the-shelf VR headsets (such as in the project with autistic young people). Therefore, developers should look to keep their XR healthcare solutions as affordable and accessible as possible if they're to encourage wider adoption in a range of clinical (or even non-clinical) settings.

- Finally, we must remember that medical data is extremely sensitive personal data. If your technology is gathering and using personal patient data, you must make users aware of this and seek their informed consent. As always, I advocate a "tread carefully" approach to collecting personal data, meaning you should only collect the data that you really need and you have to make sure it complies with legal and ethical data requirements.

Key Takeaways

In this chapter, we've learned:

- As the use cases in this chapter show, AR and VR are rightly considered breakthrough technologies in healthcare. There is enormous potential for XR technologies to improve all aspects of healthcare.

- VR and AR are already being used to improve people's well-being and enhance their exercise practices; aid the diagnosis of physical and mental conditions; enhance patient treatment (including mental health treatment, physical rehabilitation and pain management); and improve surgical processes.

- As remote patient interactions become more and more commonplace, we can expect VR and AR to play a greater role in healthcare, helping to ease the transition to remote care and making remote doctor–patient interactions more immersive and meaningful.

As we saw from the VR-enhanced exercise examples from earlier in this chapter, XR technology can help to make exercise more fun – more game-like, if you will. Which brings us neatly into the world of sporting and entertainment. Read on to discover how XR has been enthusiastically adopted by sports and entertainment providers.

Endnotes

i. Integrating Augmented Reality into Yoga Experience; Semantic Scholar; https://pdfs.semanticscholar.org/17b4/c6a1f75704e45d9e227141ef-c04978b855f2.pdf
ii. Virtual Reality can improve performance during exercise; University of Kent; https://www.kent.ac.uk/news/science/19368/virtual-reality-can-reduce-pain-and-increase-performance-during-exercise

iii. Persistent Postural Perception Dizziness (visual vertigo) project; Cardiff University; https://www.cardiff.ac.uk/psychology/research/impact/visual-vertigo-study

iv. Potential of virtual reality as a diagnostic tool for social anxiety: A pilot study; Computers in Human Behavior; https://www.sciencedirect.com/science/article/abs/pii/S074756321730417X

v. Prevalence, severity, and unmet need for treatment of mental disorders in the World Health Organization World Mental Health Surveys; PubMed.gov; https://pubmed.ncbi.nlm.nih.gov/15173149/

vi. "Mirror game" test could secure early detection of schizophrenia, study shows; University of Exeter; http://www.exeter.ac.uk/news/featurednews/title_567782_en.html

vii. Differentiation of mild cognitive impairment using an entorhinal cortex-based test of VR navigation; PubMed.gov; https://pubmed.ncbi.nlm.nih.gov/31121601/

viii. Helping kids cope with the fear of medical treatment; BBC News; https://www.bbc.com/news/business-45978891

ix. Virtual reality expands its reach; American Psychological Association; https://www.apa.org/monitor/2018/02/virtual-reality

x. You can do that?! Feasibility of virtual reality exposure therapy in the treatment of PTSD due to military sexual trauma; Journal of Anxiety Disorders; https://www.sciencedirect.com/science/article/abs/pii/S0887618517304991?via%3Dihub

xi. Virtual reality expands its reach; American Psychological Association; https://www.apa.org/monitor/2018/02/virtual-reality

xii. Mental health trust introduced virtual reality for phobia treatment; Digital Health; https://www.digitalhealth.net/2020/02/mental-health-trust-introduces-virtual-reality-for-phobia-treatment/

xiii. NHS offers new virtual reality treatment for patients with social anxiety; Digital Health; https://www.digitalhealth.net/2020/03/nhs-offers-new-virtual-reality-treatment-for-patients-with-social-anxiety/

xiv. Oxford VR; https://ovrhealth.com/how-we-can-help/

xv. How virtual reality is helping people with dementia; BBC News; https://www.bbc.com/news/av/business-49654052

xvi. Virtual Reality Technology used by Autistic Groups; West Somerset Research School; https://researchschool.org.uk/westsomerset/news/virtual-reality-technology-used-by-autistic-groups/

xvii. Cedars-Sinai Study Finds Virtual Reality Therapy Helps Decrease Pain in Hospitalized Patients; Cedars-Sinai; https://www.cedars-sinai.org/newsroom/cedars-sinai-study-finds-virtual-reality-therapy-helps-decrease-pain-in-hospitalized-patients/

xviii. Virtual Reality Emerging as Effective Pain Management Tool; Forbes; https://www.forbes.com/sites/marlamilling/2020/05/26/virtual-reality-emerging-as-effective-pain-management-tool/

xix. Recent Development of Augmented Reality in Surgery: A Review; Journal of Healthcare Engineering; https://www.hindawi.com/journals/jhe/2017/4574172/

xx. VR headsets relaxing patients during surgery at St. George's; St. George's University Hospitals; https://www.stgeorges.nhs.uk/news-item/vr-headsets-relaxing-patients-during-surgery-at-st-georges/

xxi. Vein Visualization Emerges as Premier Augmented Reality Application; AccuVein; https://www.accuvein.com/news/vein-visualization-emerges-as-premier-augmented-reality-application/

xxii. AR visualizations can help in spinal treatments, research suggests; VR360; https://virtualreality-news.net/news/2020/jun/10/ar-visualisations-can-help-in-spinal-treatments-research-suggests/

xxiii. First Augmented Reality Spine Surgery Using FDA-cleared Augmedics xvision Spine System Completed in U.S.; OrthoSpineNews; https://orthospinenews.com/2020/06/11/first-augmented-reality-spine-surgery-using-fda-cleared-augmedics-xvision-spine-system-completed-in-u-s/#:~:text=CHICAGO%2C%20June%2011%2C%202020%20%E2%80%93,surgery%20in%20the%20United%20States.

8
ENTERTAINMENT AND SPORT

Virtual reality got its big break in gaming, so it probably comes as no surprise that gaming and other forms of entertainment have enthusiastically adopted XR technologies – not just VR, but also AR. Now, these technologies are changing the wider world of entertainment and sport beyond gaming. As we'll see in this chapter, VR and AR are being used to create more immersive, engaging experiences across film, museums and galleries, theme parks, sports, music and theater, social media, and even porn. (In-car entertainment may also be ripe for a virtual makeover; German manufacturer Audi has created an experimental VR games system for car passengers that reacts to the car's movements, including acceleration, breaking and steering.)

Almost half of all investments in virtual reality fall under entertainment, showing how the industry is an intrinsic part of the XR revolution.[i] And as the technology becomes more impressive, more lifelike, we're likely to see even greater adoption of XR in entertainment – meaning there is enormous commercial opportunity in this field. Sports teams have also adopted VR and AR as a way to increase fan engagement, enhance the training process and improve sports broadcasting. There are even new e-sports that combine virtual reality games or AR graphics with real-life, physical movement – thus blurring the lines between gaming, sport and exercise.

Of course, while serious gamers may not think twice about investing in expensive VR headsets, that's not the case for everyone. That's why many of these experiences are compatible with VR- and AR-enabled smartphones, or with inexpensive VR headsets like Google Cardboard. And in the case of real-life attractions that incorporate some element of XR, the necessary kit is provided to customers when they attend. All this means XR is becoming generally more widespread and accessible for a broad range of users.

True, virtual reality won't entirely replace traditional forms of entertainment (current examples of VR rollercoaster experiences, for example, are a poor copy of the real stomach-lurching thing). But VR and AR can certainly make entertainment more immersive, engaging and accessible – and will increasingly augment more aspects of entertainment and sport.

In other words, whatever you do for fun, you'll be able to enjoy it with added VR or AR. You can watch a soccer match at home in VR, filmed using 360-degree cameras that place you wherever you fancy in the stadium. You can catch up with the news via immersive videos. You can watch a concert in VR from your own couch, or, if you attend a real-life gig, see your favorite singer augmented in AR through your phone. You can connect with friends online in a VR social media platform, and create your own little digital world. You can have a virtual movie or games night. You can immerse yourself in VR porn. You can even worship at a virtual church, run by a real-life pastor. Sound too far-fetched and futuristic? These are all current, real-world use cases that I cover in this chapter. There's no doubt in my mind that we're edging closer to a future in which the real world and the virtual worlds become inseparable, and entertainment will play a key role in this transformation.

Making Immersive Movies

If you think about the movie-making process, the director maintains tight control over every aspect of what the eventual viewer sees. Every shot is designed to focus the viewer's attention in a specific way, be it on a character's intense emotional reaction, the furtive glance between two lovers, or the killer's hand creeping out from under the bed. . . . This is how great directors provoke certain reactions, create suspense, and so on.

But in an immersive 360-degree VR film, the viewer is right in the heart of the scene, meaning they can direct their attention wherever they like – up, down, left, right, or even check out what's going on behind them. (They can't walk around in the scene, as in a computer-generated virtual environment or game, but they can turn their head in any direction.) For traditional filmmakers, this freedom to look wherever the viewer wants will pose a head-scratching challenge. Which is why I don't think we'll be seeing VR blockbusters anytime soon. However, as these use cases demonstrate, VR can still be used to enhance filmmaking and create more immersive stories.

Examples of VR films

We may not see huge Hollywood VR movies anytime soon, but there are some interesting examples of VR movies, especially short films. These are designed to create a much more immersive experience for the viewer, and potentially allow the viewer to notice new things every time they watch (because they can look all around).

Emmy-winning VR film *Invasion* is a good example. Narrated by Ethan Hawke, this animated short tells the story of two aliens who arrive on Earth planning to take over the world, but then they meet two cute little bunnies. *Ashes to Ashes* is another cool example. It tells the story of a family dealing with the grief of losing a grandfather

– and working out what to do about his last wish, to have his ashes blown up. *Ashes to Ashes* also lays bare the filmmaking process, by showing the crew filming the scenes – which is a bold and interesting move, almost making the viewer feel like they're on a movie set, rather than immersed in a story.

Using VR in the making of movies

VR is playing an increasingly important behind-the-scenes role in the movie-making process, with virtual cameras being a prime example. We all know that live action movies are filmed using cameras, but did you know that animated movies can now be filmed with "virtual cameras"? This software function, which creates lenses that act just like real cameras, is used by Pixar in its animated movies. The recent remake of *The Lion King* also used virtual cameras to film photo-realistic lions in Africa, all on an empty sound stage in the United States. With synced VR headsets, the director and crew could walk among the lions and view the scene from the first-person perspective, instead of looking at it on a monitor. Likewise, in animated movies, animators can now "walk around" inside their drawings.

Engaging with audiences through VR experiences

VR is also being used to engage with film audiences, promote movies, and provide unique experiences related to stories and characters. Disney Movies VR provides a range of immersive experiences for Disney fans, such as tagging along on red carpet events, and exploring favorite Disney scenes in more detail. Then there's IT: Float, a VR experience released to celebrate the modern remake of Stephen King's classic, which takes the user down into Penny Wise's lair.

AR, too, is proving a useful way to engage film audiences. For the 2018 Ryan Gosling movie *First Man*, which tells the story of Neil

Armstrong and the Moon landing mission, part of the marketing campaign included a web AR experience triggered by looking at the Moon. When users opened the dedicated URL on their smartphone browser, then pointed their phone at the Moon, it triggered a visually impressive experience of the Apollo 11 mission – users could even transport themselves to the Moon (by tapping on it), where they could see the American flag and spacecraft. AR promotions have also been created for, among others, *Spider-Man: Into the Spider-Verse* and *Jurassic World: Fallen Kingdom.*

Enhancing journalism through immersive films

Personally, I think we may see journalism benefit a great deal from VR films, because VR provides an opportunity to immerse users in a story and show what a situation is really like on the ground – the reality of life in a refugee camp, for instance.

This is what *The New York Times* is doing with its 360 series. Through these immersive videos, the news outlet is able to place readers at the center of a story, allowing them to look around and explore the story in more detail. One example from 2020 places viewers in the middle of the March for Our Lives protest.

More Immersive Video Games

The use of VR and AR in gaming is well established. In fact, gaming is probably the first thing people think of when it comes to XR uses. As such, there are too many examples to include them all here, but I hope the examples that I have included demonstrate the extent to which XR is penetrating the world of gaming. Adoption of XR-enhanced gaming, particularly VR gaming, is only going to grow as VR headsets get smaller and become more affordable (and as other devices, such as games consoles and PCs, become increasingly compatible with VR). We're on the cusp, then, of VR gaming going fully mainstream. Indeed, the global VR in gaming market size was valued at $11.5 billion in 2019 and is predicted to grow at an annual rate of 30 percent from 2020 to 2027.[ii] These VR games offer gamers the chance to venture "into" the game and immerse themselves in the experience through impressive 3D effects and interactive graphics. And as VR hardware, such as bodysuits and gloves, becomes more accessible, the gaming experience will get even more immersive. Meanwhile, the market for AR games, where all you need is an AR-enabled smartphone, is also expected to grow significantly over the next few years.[iii]

Examples of AR games

AR really caught the public's imagination in 2016 with the Pokémon GO game, which saw kids and adults alike using their phone to discover Pokémon characters hiding out in the real world. The Jurassic World Alive AR game does a similar thing, but with – you guessed it – dinosaurs and woolly mammoths running around your city instead. Or how about The Walking Dead: Our World, a first-person shoot-em-up-style game with zombies? Simply point your phone camera anywhere in your surroundings, watch zombies appear and then shoot them in the head. You can also snap pictures with The Walking Dead characters to share with friends on social media.

When it comes to the best-known AR games, Ingress (or Ingress Prime, as the latest version is called) has to be right up there with Pokémon GO. In the game, a mysterious new exotic matter (called XM) has been discovered around the world, and two covert factions compete to uncover the truth behind it. Ingress is a collaborative AR game, so you can team up with others in your faction to capture and defend virtual territory and discover resources.

What's great about these games is they bring a little gaming magic into the real world around us. Not fully immersive, in the way that

VR games are immersive, these AR experiences are still captivating, entertaining and – as many people can attest – thoroughly addictive. So even outside of gaming, AR features can provide a useful way to increase user engagement time within an app, game or website.

And now for some VR games

Combined with VR headsets, VR games transport users into the game environment in a way that traditional video games never could. And the range of VR games is growing all the time, from fast-paced shooter games to gentler puzzle games.

Tetris Effect, for example, is a trippy new take on the Tetris of old. It's like regular Tetris, except each level is in a new environment, complete with music and visuals that accompany the theme of that level (such as underwater noises in the underwater level). Or for horror fans, there's Resident Evil 7: Biohazard, which is played with a first-person perspective. Here, you inhabit the role of software developer Ethan Winters, survivor of a series of gruesome homicides, who is searching for his missing wife in a creepy old house. And Minecraft fans can indulge themselves with Minecraft VR, which is just like regular Minecraft, except you're fully immersed in your world as you build and fight stuff.

The long-running video game franchise Ace Combat has also released a VR version with Ace Combat 7, which uses detailed aircraft and photorealistic scenery to make players feel as if they're really flying in a fighter jet. And sticking with the flying theme, there's the enduringly popular Microsoft Flight Simulator. In 2020, Microsoft announced the program would soon be supporting VR headsets (beta testing of this was under way at the time of writing) to give users a more immersive experience of flying a plane. People have been raving about this home flight simulator for years, so the prospect of being able to use it with a VR headset is really exciting.

EXTENDED REALITY IN PRACTICE

You can even play poker in VR with the PokerStars VR Poker Tour, a free-to-play poker tournament that welcomes players into a virtual poker environment, where you play against avatars of other competitors. This could pave the way for a range of competitions and tournaments to go virtual in future.

Virtual Visitor Attractions

In Chapter 6, we explored how VR is opening up new educational experiences through virtual excursions and adventures. Let's look at some more examples of virtual attractions that have been designed to immerse visitors in an experience, without having to leave home.

Virtual exhibitions, galleries and museums

Many museums and galleries around the world are now creating virtual tours to bring their exhibits to life. These vary in complexity – some are simple 360-degree walkthrough videos, while others have more interactive features for various exhibits – but all are designed to increase engagement and bring collections to life.

The National Gallery in London has created three virtual tours that allow visitors to explore the gallery, via their desktop, mobile or VR headset. There's a VR tour of the Sainsbury Wing, for example, which shows off the gallery's Early Renaissance paintings. Likewise, the J. Paul Getty Museum in Los Angeles has created a virtual tour, where you can look around gallery spaces and click on artworks for more detail. Or there's a virtual tour of the Vatican Museums in Rome, complete with its murals, tapestries and richly decorated vaulted ceilings. In other words, you can see the Sistine Chapel in vibrant, 360-degree detail, without having to queue for the privilege. The Louvre also has a VR experience called "Mona Lisa: Beyond the

Glass." Complete with interactive design, sound and animation, the downloadable experience lets users learn more about the painting, its texture and how it has changed over time.

The year 2020 saw the launch of VOMA, the Virtual Online Museum of Art, the world's first completely virtual museum. With art loaned from leading galleries around the world, VOMA presents curated exhibitions, in a beautiful virtual environment that's completely free to download. VOMA plans to include plenty of new features and artist projects over time, so the idea is users will see their copy of the museum update.

What's great about these experiences is that – as well as being able to explore museums and galleries from wherever you are in the world – you can, in many cases, really get up close and inspect the artworks and exhibitions in minute detail. Anyone who's ever seen the *Mona Lisa* in real life, with its protective glass and selfie-snapping crowds, will know that's not an easy thing to do. VR can also help to demonstrate the true scale and majesty of exhibits in a way that 2D images can't. Yes, it's a different experience to the real-life pleasure of wandering around a gallery or museum, but as a way to bring art and culture to life, these virtual experiences are brilliant.

Virtual zoos

Virtual zoos promise to bring users closer to the animals than real-life zoos, potentially allowing them to interact with animals in an immersive, safe way. One example comes from Zoo World VR, which was released in 2020. This VR experience, by Intentio Education Game Studios takes users to different parts of the world so they can see and learn about the different native species. Combining the zoo experience with gaming, users help out animals on a series of environmental-themed quests.

Like real-life museums, real-life zoos are also creating their own VR and AR experiences. Chester Zoo in England has created an AR app called Wilderverse that transforms your home environment into a jungle, so you can interact with apes and take part in conservation challenges. And during the coronavirus pandemic, when the zoo was closed to visitors, Chester Zoo also created a series of "virtual zoo days," where the zoo went live on social media and YouTube for a whole day of fun activities.

Virtual theme parks

That's right, you can now put on a VR headset and experience theme park rides without getting up off the couch. The VR Theme Park Rides experience by Mexico-based software developer EnsenaSoft is a collection of three theme parks in one, with 12 rides in total – from gentle teacups and Ferris wheel rides to a classic rollercoaster. And you don't have to wait in line for tickets, rides or hot dogs. In fact, there are a number of different VR rollercoaster apps, games and experiences, including RollerCoaster VR Universe, which has roller-coaster rides that rocket through a range of different environments, including outer space and under water.

Making Real-World Attractions More Immersive

Just as you can now go on a virtual theme park ride from home, you can also enjoy a VR-enhanced experience at real-life theme parks. In other words, you attend a physical theme park or attraction, where VR is used to spice up the entertainment and improve the visitor experience. In some cases, VR is the main attraction itself, not an add-on. Let's see what all this means in practice.

VR as the main draw

The Vertigo VR entertainment center in my home town of Milton Keynes is my son's current birthday destination of choice. Described as a "cybersports hub," there's a VR arena where you can play VR games, and VR motion pods that use motion, sounds and wind to make you feel like you're soaring through the sky (among other things). There's also a 5D VR cinema with moving chairs and real effects such as water and smoke.

Elsewhere, theme parks dedicated entirely to VR have opened, such as VR Star Theme Park, located in southwestern China, or Dubai's VR Park, which describes itself as the world's largest VR park. Both offer a huge variety of VR rides, games and experiences. In the future, we'll see more and more of these VR entertainment spaces, where extra dimensions are used to make the entertainment more fun and immersive.

VR-enhanced rides

Traditional theme parks are now beginning to incorporate an element of VR into some of their rides, usually by overlaying a VR experience on top of an existing rollercoaster, drop tower or water slide. In one example, riders on the Kraken Unleashed rollercoaster at SeaWorld in Orlando, Florida are given VR headsets to enhance the traditional coaster experience. So, as well as all the usual loops, twists and stomach-dropping thrills, the VR headset takes riders into a digital underwater world that matches the theme of the ride.

Nowadays, we even have VR water slides. Yes, you read that correctly: we're now able to combine sensitive technology with a whole lot of water. The Therme Erding water slide park in Germany has a VR slide that, thanks to custom-made waterproof VR headsets, transports riders to different virtual environments as they glide down and around the chute.

These examples may sound a bit bonkers, but adding VR to existing rides makes good business sense when you think about it. It gives parks a new way to attract visitors, without the huge cost associated with installing a new ride. However, adding a VR element to rides isn't without its challenges. For one thing, distributing, collecting and cleaning headsets is a drain on staff time, and wait times can increase because of the extra time it takes for riders to strap on and adjust their headset. Then there can be technical challenges when the VR visuals don't sync up exactly with the physical ride. But, done well, these experiences can blend the virtual and real worlds in an exciting new way to thrill visitors – and keep them coming back for new, easily updated experiences.

Hado

While we're on the subject of blending real-life and virtual entertainment, I can't neglect HADO, the world's first physical e-sport. One of my favorite examples in the entire book, HADO is essentially dodgeball, but with an added AR element. Now, I've always loved dodgeball, but this takes it to the next level by equipping players on a real-life court with virtual energy balls and shields.

In other words, you go to a physical HADO arena (at the time of writing, there are 60 HADO arenas in 15 countries), and have the best game of dodgeball in your life! You can play one-on-one, two-on-two, or three-on-three and the goal is always the same: to hit the other team or player with as many energy balls as you can to drain them of their "lives" and earn points during each timed event. You can defend yourself with shields, which weaken if they get hit by the opposing team's energy balls. There are even HADO competitions and tournaments, including the HADO World Cup, and a Pro HADO League.

What's really clever about this is players have all the freedom of movement they would in a regular game of dodgeball, meaning they don't have to be tethered to a gaming system, controllers or cables; all they need is a headset and a small motion sensor worn on the wrist or forearm. HADO – which was created by Japanese startup Meleap – may be the first physical e-sport created, but I'm certain it won't be the last. The merging of physical athletics with AR (or even VR) technology to create immersive new sports could be a big growth area in future.

Which brings us neatly on to the world of VR and AR in sports.

Use of XR in Sports

XR technologies are also infiltrating many aspects of sports, from fan engagement and the spectating experience, to referee-type decision making in matches and athlete training (which I touched on briefly in Chapter 6). In fact, as we'll see in this section, VR and AR are now widely used in a range of sports.

In-match technology

If you've ever watched tennis – I'm a huge Wimbledon fan – you'll be familiar with Hawk-Eye technology, which provides a 3D representation of a ball's trajectory to ensure line judge decisions are fair. Not only does this make for a fair game, it also adds an extra frisson of excitement and suspense to a match – I speak from experience when I say fans love it when a decision is referred to Hawk-Eye for confirmation. The technology can also be used to overlay virtual graphics onscreen to create, for example, a green screen analysis of a player's backhand form.

In pretty much any kind of sport, AR can be used to overlay digital representations of ball trajectories, tactics and scenarios to explain what just happened on the field or preempt what might happen next. Indeed, pundits and commentators have made rudimentary use of this sort of thing for years, with simple arrows and graphics projected onto footage of the real-life action.

You know the yellow line that's projected onto the field in televised American football matches? That's AR in action. Marking the first-down line, this magic yellow line makes it easier for at-home viewers to follow the action and work out which plays are successful, without having to wait for the referee or commentator to confirm. It was developed by Sportvision, leaders in sports broadcasting technology, and first used in 1998, making it one of the first high-profile uses of AR. Now, viewers are so used to seeing that yellow line onscreen, it can make the experience of going to a real-life match (where there's no yellow line on the actual field) pretty weird.

Improving fan engagement and the viewing experience

Watching a match at home may not be as atmospheric as watching in a stadium but with some added interactive virtual elements, viewers can certainly enjoy the experience a lot more than they used to. Therefore, as well as making it easier to follow the action (as in the yellow line example I just mentioned), VR and AR can also help to deepen fan engagement.

In soccer (or football as we call it in the UK), teams are beginning to experiment with 360-degree cameras, giving home viewers the chance to watch the action from different locations in the stadium, as if they were really there. For example, customers of BT Sport, part of the UK's British Telecoms company, can now watch selected live Premier League soccer matches and highlights in 360-degree VR through the BT Sport app. This creates a unique viewing experience because, rather than being stuck with fixed-camera positions, you can look around from different viewpoints in the stadium. In the future, you may even be able to put yourself on the pitch as one of the players – for example, to watch a penalty kick from the position of the goalkeeper. I firmly believe this is the future of sports viewing.

Similarly, in basketball, the NBA has a Magic Leap app that lets fans watch live matches, replays and highlights in AR and access interactive features such as team and player stat comparisons. According to the NBA, the initiative is designed to make sports more interactive and to attract more younger fans. The NBA also partners with TNT and Intel's True View VR technology to create VR broadcasts of basketball matches – fans with compatible headsets simply download the NBA on the TNT VR app to begin watching.

Elsewhere, AR can be used to let fans digitally attend matches that are played behind closed doors. During the coronavirus pandemic, when attending a soccer match in person was out of the question, playing in silent, empty stadiums must have been weird for players. But London firm OZ Sports believes its AR and AI technology has the answer. Its OZ Arena product lets fans "appear" at a match, digitally speaking. Teams and broadcasters can either overlay avatars of fans onto empty seats or select an audio-only option, which broadcasts real-time fan audio to create a noisy crowd effect.

And for fans who are lucky enough to be there in person, the in-stadium experience can also be enhanced through XR. In one example, the Dallas Cowboys football team created an AR "Pose with the Pros" experience, which could be activated by fans at the stadium. It's basically a huge interactive screen that lets fans select up to five of their favorite Dallas Cowboys players, then grab a group photo with them. After debuting the AR experience, the team reported that it had generated more than 50 million social media impressions.[iv]

Improving training practices

Both AR and VR can be used to create custom training practices. Training data can be visualized in interesting new ways with AR, and realistic training exercises and scenarios (and even opponents) can be simulated with VR.

In fact, many sports teams are investing millions in XR technologies to improve and diversify athletes' training practices, and this is particularly prevalent in American football. Notable NFL and college football teams, such as the Dallas Cowboys, New England Patriots, Tampa Bay Buccaneers and Vanderbilt University are using VR in training, and reaping the benefits of players being able to train without having to set foot on a football field. VR can be used to simulate actual game conditions (even actual opponents), helping to making training sessions more effective.

Founded in 2013, EON Sports is one of the most successful VR developers working in the field of sports training applications. Those football teams I just mentioned? They all use EON Sports' technology to create realistic training simulations. These simulations allow players to practice against virtual opponents, who are programmed to replicate the playing style of real-life opponents. It can essentially be used to create a trial football match against upcoming opponents in the days before a real-life match. Simulated training technology like this combines the use of VR head-mounted displays with physical sports equipment (such as a ball, safety equipment, or, in the case of baseball, a bat).

New sports and new competitions?

As we've seen so far, VR and AR can be used to bring fans closer to the action. But it may also bring about new e-sports (as in the case of HADO, which I've already mentioned) and new forms of competition – even allowing fans to compete against their favorite pros.

One brilliant example of this comes from Formula E's Live Ghost Racing experience, where fans can pitch themselves against real-life Formula E racing drivers live, in real time, as they race on virtual versions of streets around the world. Or there's the Echo Arena VR game, a multi-player zero-gravity game that's a bit like a mashup of football and Quidditch (minus the broomsticks). With such e-sports, players are playing a virtual game, yet there's a lot of physicality involved – creating a sort of hybrid of sports, exercise and gaming that could really take off in future.

XR in Music and the Performing Arts

Artists and entertainment companies are having to find new ways to make live festivals and concerts worth the entrance fee, and to provide the unique experiences fans demand. VR and AR can add value to performances in a number of ways, as these experiences show.

XR-heightened live performances

U2 fans who attended the band's Experience + Innocence tour were given an AR-infused treat via an 80-foot screen on stage – fans who pointed their smartphone cameras at the screen saw a huge projection of Bono outlined in electric blue. Maroon 5 did something slightly different by providing a live AR karaoke experience at a limited number of concerts – where fans could use a special Snapchat lens to film themselves singing along during the show. And for

Eminem's Coachella set, fans could download an app to view the rapper on stage accompanied by images that matched the songs.

The use of AR in live music performances is still in its early days, but these examples show how AR can help artists build buzz around events, build excitement during a gig and create deeper connections with the crowd. Think of it as the twenty-first-century version of crowd-surfing, if you will.

Some theatres have also been experimenting with using VR and AR to bring an extra dimension to live theatre performances. One interesting example comes from SOMNAI, an award-winning immersive theatre experience that combined VR with live actors on a physical set. Situated in a 20,000 square foot warehouse, the "lucid dreaming" experience – which even included pajamas for attendees to put on – immersed the audience in VR dream sequences complete with multi-sensory elements (such as audio) and actors playing dream guides.

A lower-tech example comes from my family's 3D experience at a Christmas pantomime performance in Milton Keynes. Everyone in the audience was given 3D glasses, and at one point in the show we were invited to put on the glasses and go on a little 3D ride. Although it was nothing like a slick VR experience, it certainly added an engaging extra dimension to live theatre. And my kids loved it. In the future, virtual elements could be added to all sorts of theatre experiences – or even allow audiences to enjoy performances from home (more on that coming up).

Then there are those holographic performances that bring artists such as Whitney Houston and Buddy Holly back from the dead. Many may see this as just a fad (indeed, I'm not convinced there's a huge future in holographic performances). Others find the idea vaguely creepy or disrespectful. But it does highlight how virtual performances – where the artist doesn't have to be in the stadium, or even alive – are possible these days. Which brings me on to . . .

Virtual gigs

So we've seen how XR can be used to enhance gigs on real-life stages. But how about attending a virtual gig? It's possible, thanks to VR.

In one high-profile example, the ever-popular Fortnite game (which I mentioned in Chapter 5) has begun hosting virtual gigs on the platform, attended by millions of players watching within the game. Electronic DJ Marshmello and superstar rapper Travis Scott have both performed virtual gigs in Fortnite so far, and we can expect the platform to host many more in future. In the days before Scott's performance, players in the game could see a stage being constructed, with inflatable Travis Scott heads around it – which no doubt helped to build buzz in advance of the gig. Then, when the show started, a giant Scott stomped around the game's island, accompanied by delighted players. Visuals changed to accompany different tracks (for example, transforming Scott into a cyborg), and at one point the whole crowd was submerged underwater. It was only a short, 15-minute set, but it was certainly popular – a record-breaking 12.3 million players attended the virtual gig,[v] which is pretty mind-blowing.

There's also MelodyVR, which creates immersive VR music performances for fans via its app. The platform hosted a free "Live from LA" concert series during the pandemic, staged at an empty Los Angeles studio and headlined by John Legend. Viewers could virtually attend the concert using their VR headset or smartphone and feel like they were in the same room as Legend. In fact, when the coronavirus pandemic started, MelodyVR hosted 10 times as many concerts as it had in the same time frame the year before, and experienced a

1,000 percent increase in app downloads,[vi] indicating that audience appetite for virtual live-streamed performances is certainly there.

French electronic music composer Jean-Michel Jarre has also been a leading champion of VR concerts. At one live-streamed concert in 2020, fans who were watching via a VR headset could interact with each other through their virtual avatars. And, to enhance Jarre's usual trippy beats and light show, attendees could take virtual "pills" that made the screen change color. That's right, you can even get digital drugs at your digital concert.

As live-streamed concerts become a more regular feature (pandemic or no), we can expect artists to use VR as another way to connect with audiences and make the experience more enjoyable and engaging for fans.

And Now for Something Completely Different . . . Adult Entertainment

Whatever your opinion of porn, there's no denying the adult entertainment industry is big business (estimated at $35 billion in 2019).[vii] It would be remiss to write about XR in entertainment and not talk about adult entertainment. So, brace yourself; let's see how VR in particular is being enthusiastically adopted by the industry and users alike.

VR porn and virtual sex

Head to PornHub (well, maybe not on your work computer) and you'll see that there's a dedicated, fast-growing category for VR porn. But what exactly is VR porn? With VR goggles, you can immerse yourself in 3D, 360-degree porn, set in a range of different environments (in the woods, in a public setting such as a restaurant, you name it). So, instead of staring at a screen, you can be transported to another place entirely.

There's also a range of hardware designed to enhance the experience of watching porn in VR, such as the Titan VR by KIIROO. This interactive vibrating stroker literally brings another dimension to VR porn experiences by adding the sensation of touch. What's really interesting about this is it can connect with other online devices, meaning you can actually feel what your virtual partner does via your own device – hence, VR sex. Or it can be used, well, solo, in conjunction with thousands of videos that can be synced up to the stroking device. There's also Virtual Mate, which combines a masturbator sleeve with a photo-realistic lover that users can look at on-screen via their smartphone or VR headset. The virtual lover is programmed to sync up with the stroking sleeve, so it reacts to the user's movements and speed.

It's even possible to have VR-enhanced sex with silicon women, in the real-life setting of a brothel. At the Naughty Harbour brothel in the Czech Republic, customers put on a VR headset that plays video of their favorite porn star while the user has sex with a silicon love doll.

The future of sex?

As I mentioned in Chapter 3, VR sex isn't without its, if you'll forgive the expression, sticking points. For one thing, there's the possibility that users could have VR sex with digital versions of real-life people without their knowledge, which is a pretty creepy thought.

VR sex may also impact our real-life relationships, as we work to redefine what's acceptable and what's not with partners. For example, someone may be completely fine with their partner viewing regular online porn, but what about immersing themselves in VR porn and having virtual sex with another partner online, complete with the sensation of touch? Will active participation be considered cheating? The notion of what it means to be faithful (or not) could be tested, and couples may find themselves having to draw boundaries that previous generations of couples couldn't have imagined.

Getting Social with XR

For many people, the brief bit of entertainment time we get in our busy lives is spent on social media. So perhaps it's no wonder many companies are applying VR to the world of social media to create "social VR" platforms, where users can hang out, play games and socialize with avatars of their mates in a collaborative, digital space. If this sounds familiar, it's the idea behind Facebook's Horizon social VR platform, which I mentioned in Chapter 4. But Facebook isn't the only company getting in on the VR act.

Hanging out in social spaces

Let's look at a few examples of other social VR platforms. VRChat is a VR social platform that lets users play, hang out and chat with each other using 3D audio and lip-synced avatars. Users have the power to create their own social worlds, and there are more than 25,000 community-created worlds and counting on VRChat. Another example, called Bigscreen, aims to replicate the experience of watching a movie with mates, even when you're all in different locations. Through Bigscreen you can hang out at virtual movie nights, parties and games nights with up to 12 of your friends from all over the world, and the platform is compatible with over 50 streaming TV channels.

AltspaceVR lets users attend free live events with DJs, comedians, authors and more, all from home, and with your friends in their own home. There are also interactive games to play on the platform. Or there's Sensorium Galaxy, an ambitious social platform that attempts to create an all-digital alternate universe. In fact, with Sensorium, clubbers can put on a headset and dance the night away with fellow clubbers in cyberspace, complete with a live DJ. The company has partnered with the team behind some of Ibiza's biggest nightclub venues to create a state-of-the-art virtual club within Sensorium Galaxy that's dedicated to electronic dance music.

You can also recreate your favorite bar or pub in VR. Missing his local pub during lockdown, British guy Tristan Cross re-created it in VR. First, he had to learn how to build 3D models, which he learned via YouTube, then it took him two-and-a-half weeks to create the impressive virtual pub.[viii] He even recorded a call with his friends, and re-created it with animated digital avatars in the virtual pub.

Virtual worship

If you can hang out in a virtual spaces, why not attend virtual church? Pastor D.J. Soto is putting that idea into practice with his own VR mega-church, which allows members to worship in a virtual space. Inspired by visiting the AltspaceVR platform back in 2016, Soto quit his job as a pastor at his local church and set about creating his own radically inclusive virtual church. The result, VR Church – which worshipers can attend on Altspace – is the first house of worship to exist only in VR.[ix]

Soto believes virtual churches are the way forward for churches beset by slumping attendance numbers, especially as headsets become more affordable, or as a way to welcome worshippers who might otherwise be alienated by traditional churches. It's an interesting idea, that you could watch VR porn or have VR sex one day, then attend VR church the next, but this demonstrates how XR will eventually make its way into many aspects of our everyday lives.

Looking ahead, I believe social VR platforms will extend far beyond straight-up social spaces to become platforms that house an entirely digital alternate world, where users can attend concerts and night-clubs, watch their favorite sports team, play a game of zero-gravity football (or another e-sport), catch up with friends, and even have sex. Much like in the movie *Ready Player One*, we may find ourselves strapping on a VR headset in our spare time and entering a world where we can indulge all of our hobbies, social activities and interactions in one digital space – a virtual world where we can be whomever we want and do whatever we want. That may be the future of entertainment.

Lessons We Can Learn from Entertainment and Sport

Regardless of whether a business operates in the sports and entertainment space, there are still some interesting lessons business leaders can learn from these use cases. Key learning points for me include:

- It's clear that XR provides a new way to build additional content around key products. Think of how Disney's VR experiences build on Disney stories and characters, or how news outlets can enhance their journalism through immersive video, or how theme parks can layer VR on top of traditional rides. In this way, XR doesn't have to be your main product or attraction – more, it's a way to add extra value for customers and keep them engaged for longer.

- Don't overlook the potential for VR or AR to enhance your internal business processes. I'm thinking here of the film directors and animators who use VR to "walk through" scenes, rather than viewing them as flat, 2D images onscreen. VR could, for example, enable technicians and engineers to walk through their systems virtually before they carry out repairs or maintenance. Or, as we saw in Chapter 6, you can create immersive employee training and education experiences.

- Remember that XR isn't just about creating a digital experience that's separate from the real world. In fact, it can make real-world experiences more engaging and enjoyable, as in the examples of theme parks and HADO arenas. Pretty much any real-world customer experience could, in theory, be enhanced through AR or VR – including shopping, hanging out in a bar with friends (remember the VR cocktail from Chapter 5?), going to the theatre, you name it.

Key Takeaways

In this chapter, we've learned:

- VR first gained popularity in gaming, and has since been widely adopted across pretty much all forms of entertainment and sport, including film, museums and galleries, theme parks, music and theatre, social media, and even porn.

- Entertainment has seen huge investment in VR technology, making it a key part of the XR revolution and an enticing business opportunity.

- While VR and AR experiences won't entirely replace traditional forms of sports and entertainment, they are certainly making them more immersive, engaging and accessible. So, whatever you do for fun, there's likely to be an AR- or VR-enhanced version available or coming your way soon.

- What's more, entirely new formats are being created thanks to XR, such as immersive theatre experiences that combine real-life actors with VR video, or e-sports that combine VR or AR graphics with real-life physical movement.

Sectors such as gaming and filmmaking may be used to incorporating new technologies all the time. But what about seemingly more traditional industries like real estate and construction? As we'll see in the next chapter, these are also embracing XR technologies.

Endnotes

i. Virtual Reality for the Entertainment Market; Jasoren; https://jasoren.com/virtual-reality-for-the-entertainment/
ii. Virtual Reality in Gaming Market Size, Share & Trends Analysis Report; Grand View Research; https://www.grandviewresearch.com/industry-analysis/virtual-reality-in-gaming-market
iii. Augmented Reality Mobile Games Comprehensive Study by Application; Advance Market Analytics; https://www.advance-marketanalytics.com/sample-report/101677-global-augmented-reality-mobile-games-market
iv. Fans Excited to "Pose with the Pros"; Dallas Cowboys; https://www.dallascowboys.com/news/fans-excited-to-pose-with-the-pros
v. More than 12 million people attended Travis Scott's Fortnite concert; The Verge; https://www.theverge.com/2020/4/23/21233946/travis-scott-fortnite-concert-astronomical-record-breaking-player-count#:~:text=Share%20All%20sharing%20options%20for,attended%20Travis%20Scott's%20Fortnite%20concert&text=Travis%20Scott's%20first%20virtual%20performance,it%20also%20broke%20a%20record.
vi. Napster's New Bosses Want to Make a New Kind of Music-Streaming Giant; Rolling Stone; https://www.rollingstone.com/pro/news/melodyvr-buys-napster-streaming-service-1049716/
vii. Global Online Porn Market Was Estimated to Be US $35.17 Billion in 2019 and Is Expected to Grow at a CAGR of 15.12% Over the Forecast Period, Says Absolute Market Insights; Cision PR Newswire; https://www.prnewswire.com/news-releases/covid-19-update-global-online-porn-market-was-estimated-to-be-us-35-17-billion-in-2019-and-is-expected-to-grow-at-a-cagr-of-15-12-over-the-forecast-period-says-absolute-markets-insights-301043437.html
viii. Why I recreated my local pub in virtual reality; BBC News; https://www.bbc.com/news/av/technology-52833546
ix. This Pastor Is Putting His Faith in a Virtual Reality Church; Wired; https://www.wired.com/story/virtual-reality-church/

9
REAL ESTATE AND CONSTRUCTION

Although I talked about gaming in the last chapter, let me start this chapter with a brief gaming example. Just when you think there's a game for everything, along comes Landlord GO (inspired by Pokémon GO perhaps?), an AR-based game that lets users buy, sell and collect rent on real-world properties and landmarks in their surroundings. (The game's makers, Reality, say it's the first real-world AR game that uses real buildings and real prices to create a realistic, data-driven property game.)

This brings us neatly on to the use of XR technologies in the world of real estate, spanning everything from marketing and selling properties, to designing and constructing buildings.

Particularly for real estate agents, VR is proving to be a game-changer, largely because it allows buyers to view properties from a distance, via immersive 3D tours. And even when clients don't have their own headsets, agents can invite them to enjoy virtual tours from the office – allowing clients to view multiple properties without having to drive all over town. (As VR headsets become more ubiquitous, I predict more and more of us will be going on virtual house tours from the comfort of our own sofa.) These virtual tours save realtors and their clients a lot of time. And, due to their immersive nature, they also help to create a deeper emotional connection than simply looking at photos or floor plans, and provide a better understand of the property. Even better, clients can spend as long as they like in the virtual version of the property, or visit it as many times as they like, without driving their agent nuts. In other words, VR helps to alleviate many of the pain points associated with house hunting – all that time spent going around unsuitable properties, all those miles, trying to make sense of floor plans – and turn the house-hunting process into something much more engaging, immersive and interactive. Without being a drain on the agent's time.

As we'll see in this chapter, the larger real estate agencies are already embracing VR tours to improve the customer experience and increase efficiency, and this will no doubt filter down to smaller providers in time. (All you really need to create a virtual tour is panoramic photos of the property. These can then be turned into a virtual tour using something like Google VR's Tour Creator, or by partnering with a VR tour specialist.)

AR is also proving useful in real estate, particularly when it comes to the virtual staging of properties or providing immersive virtual instructions for tenants – again, uses that are designed to improve the customer experience and generally make the whole experience of buying or renting a property easier. In the future, we may even see "digital real estate agents" that guide us around a property via AR glasses, when the listing agent isn't available to conduct a tour. Visuals and text could pop up in the buyer's field of view to tell them all about the brand-new marble countertops or the state-of-the-art sound system, so they get to explore the property on their own, without missing out on all the selling points.

Architecture and construction are also beginning to use XR technologies more widely. In architecture, VR is bringing 3D models to life and giving clients a much more immersive view of their project – making the approval process far easier, and allowing clients to have greater input into the design (because they can "step inside" their building long before it is built).

And on construction sites, AR is helping to improve accuracy, make sites safer and streamline the inspection process. It may seem odd to picture builders and electricians walking around with AR headsets on, but remember that AR is rooted in reality and doesn't obscure the real-world view; it enhances it. This makes it much more suited to use on a busy construction site, as opposed to VR, which creates a completely separate environment. (VR is better suited to training and education for construction workers – for example, immersive safety training.)

I predict VR and AR will play an increasingly important role in these industries, and not just for big real estate agencies and on huge construction sites, but across a range of independent firms and smaller construction projects. Just as with the internet or 3D building modelling, XR technologies will likely become an integral part of workflows in the future. But for now, let's look at some current real-world uses from the world of real estate and construction.

XR in Real Estate

Let's explore how XR – especially VR – is enabling the real estate sector to provide a slicker, easier and more immersive experience for clients.

Savills – providing guided visits and interactive property tours

When you're buying a house, you'll usually visit several properties before deciding which one is right for you. It's a time-consuming process – for you and for the real estate agents involved. And if you happen to be searching for property in an area that's far away from where you currently live, it's even more complicated.

VR is helping to solve this problem through virtual property tours, which allow people to visit properties for sale, virtually, and immerse themselves in a simulated 3D walkthrough that perfectly matches the real-life property. It's easy to see how this makes viewings more efficient for both house hunters and realtors.

Generally, these virtual property tours fall into two camps. There are guided VR tours, where the video guides the viewer around the property in a specific order (much like a standard promotional video). These can be in the form of relatively simple 360-degree videos, or full VR experiences that are best viewed through a VR headset. Then there are interactive visits, which allow the viewer to choose which parts of the property they want to explore, in what order – by selecting certain "hotspots" within the field of view, the user can decide where they move to next. Naturally, these interactive experiences are more expensive and complex to produce, so they're currently less common than guided VR tours.

In Chapter 5, I briefly mentioned how luxury providers like Christie's are already using VR in this way. In the UK, leading property advisers Savills became one of the first real estate agents to use VR to market property when they created a slick VR experience to market a multi-million-pound mansion. After the property had been scanned, filmed and photographed, both guided and interactive VR tours were created to allow viewers to explore the property in detail. The virtual experience even included audio, for example, of a roaring fireplace in a room and birds singing in the garden. Of course, most of us would still go to view a property in person before buying it, but tools like this allow us to spend that time only on the properties that we're really excited about, instead of schlepping around dozens of places.

In addition, Savills has been using virtual tours to market new homes – which is another great way to use VR since viewers can tour virtual versions of properties that haven't been built yet. Far more compelling than looking at 2D plans and images, virtual tours like these could revolutionize off-plan or sight-unseen sales.

Virtual staging of properties

If you've ever viewed a completely empty property, you'll know how difficult it is to get an accurate sense of the true scale of rooms, or how your furniture might fit. That's why agents generally prefer to show properties furnished – it's just easier to sell properties that way. However, showing properties as furnished isn't always possible when a home is currently unoccupied, not unless the agent rents furniture to "stage" the space.

This is where virtual staging comes into its own. Now, there are companies such as roOomy and BoxBrownie who will digitally augment photos of an empty property and fill it with furniture to suit various different styles, such as industrial, farmhouse, or contemporary. They'll even stage the garden with attractive patio furniture. This also opens up the possibility of realtors staging properties for photos and VR tours according to each individual client's tastes. In the first instance, this would probably be reserved for high-roller clients, but there's no reason it couldn't be applied to all clients – for example, when showing properties to a young family, you could digitally fill the images with a crib, toys and a changing station. Or for a young, single guy, it could be decked out with modern furniture and the latest gadgets. It all helps to sell the lifestyle associated with a property and how it would feel to live there.

Luxury real estate brand Sotheby's International Realty has created its own AR app to digitally furnish an empty room with beautiful furniture (which can also be purchased directly through the app). Called Curate, and developed in partnership with RoOomy, the app is useful for individuals with deep pockets who are looking for some interior design inspiration, but it's also been used by Sotheby's real estate agents to stage empty properties and demonstrate what size furniture will fit in different rooms.

You might remember I mentioned Matterport's spatial scanning software in Chapter 4, which can be used to create incredibly accurate 3D digital walkthroughs of any property. Now, Matterport has partnered with UK-based VR staging provider VRPM to enable Matterport's 3D models to be virtually staged, creating beautifully furnished residential spaces or commercial interiors. UK housebuilder Mulberry Homes is an early adopter of the Matterport/VRPM service.

Much like the IKEA app that I mentioned in Chapter 5, this virtual staging idea is so simple, yet incredibly compelling and effective. I'm excited to see how it might develop in future. For example, we could put on AR glasses at in-person property viewings and superimpose our own furniture into homes to see how it would really look.

Kingdom Housing Association and XR-enabled property management

Landlords and real estate agents who handle rental properties spend a lot of time managing tenants, answering their questions and dealing with routine maintenance issues. But there's a big difference between simple queries like "How do I get the thermostat to work?" and major issues such as "Help! The boiler has erupted!"

Scottish housing provider Kingdom Housing Association is using AR to sort the easy fixes from the things that absolutely require a visit to the property. Implemented just before the coronavirus lockdown (incredible timing, considering social distancing meant visiting properties for routine issues was out of the question), the technology allows trades operatives to guide tenants through simple, routine repairs. Essentially, tenants can have a tradesperson virtually visit their home and, with the help of a smartphone or tablet, the tradesperson can view the problem in real time and guide the tenant on how to fix it. The solution is a partnership between Kingdom Housing Association, housing consultancy DtL Creative and Swedish-based remote guidance software specialists XMReality. The housing association reported a positive response from tenants,[i] so we may see more uses like this in future. In fact, a range of routine repairs and inspections could be carried out with the help of AR software.

Alternatively, landlords could create VR experiences that give detailed virtual instructions to tenants, for example, on how the utilities services work (so instead of phoning the landlord, you put on a VR headset or download an app and watch a brief tour in immersive 3D video). This obviously takes time and money to create but could save a lot of effort in the long run – making it potentially a very useful option for housing associations who manage a lot of similar properties or where there's a high turnover of tenants (for example, in vacation properties).

XR for Architects and Their Clients

Just as VR makes selling houses that little bit easier, it can also make designing them – and collaborating with clients – easier. As such, we're beginning to see more and more architects use VR to enhance the design process, and also to give clients an immersive experience showing exactly how a space will look. In time, it could transform the process of designing buildings and sharing concepts with clients.

Immersive collaboration

With VR, design teams can create and experience ideas together and collaborate from wherever they are in the world. That's the goal of immersive collaboration specialists The Wild, whose software creates a shared virtual space for architects and designers. In other words, an architect can see what their colleague is designing and vice versa, experience designs at scale and even meet up together virtually. The tool is compatible with 3D modeling and design software such as Revit and SketchUp, so architects can import their existing models.

I find this use case interesting because it could be applied across many different industries, not just architecture and design. If you think about it, successful collaboration relies on people's ability to express ideas, view others' perspectives and communicate seamlessly. Tools like The Wild could help a great deal because they bring ideas to life in real time and allow dispersed teams to collaborate more easily.

Viewing 3D models in VR

Architects have been designing with 3D models for many years, but the incorporation of VR allows those models to be seen in a whole new way, so that the team and clients can experience the design in much more detail. Using software like Enscape or IrisVR, architects can turn their, for example, SketchUp model into an immersive VR experience. This allows architects and their clients to understand designs better (more on that coming up next) – and also to test different options at full scale early in the design process to see what really works and what doesn't. In this way, VR could help to boost innovation and encourage architects to experiment with new designs.

Visit your new home before it's built

I've mentioned already how VR helps architects communicate their concepts to clients in a more immersive way (see also the Urbanist

Architecture example mentioned in Chapter 5). Let's explore this concept in a little more detail.

For clients who are having a new home (or commercial property) built, or having extensive remodeling work done (say, a large extension), VR means they can experience their new space before it is actually built. They can see the design in much more detail and make changes at an earlier stage, which saves time and money – because making changes once the build is under way can have a huge impact on the schedule and costs.

Using VR with clients is helpful because not everyone can look at a plan or 3D model and visualize what it's really going to look like when you step inside. With VR, clients can do just that – step inside their design, and virtually "test drive" the space. As an example, my wife and I had our kitchen and bathroom remodeled recently. It was a pretty big, expensive job and we wanted to make sure we got it right. However, while I find it fairly easy to visualize designs, my wife doesn't. She needed to "see" what the spaces would really look like from the inside before agreeing to a decision.

This all benefits the architect, too. For one thing, they can increase client satisfaction by ultimately creating designs that better suit the client (because the client will have been able to experience the space and make informed decisions earlier in the process). VR also makes it easier for clients to sign off on a project with confidence, hopefully speeding up the approval process. Plus, VR can give the architect a greater insight into the client's priorities and how they would really use a space. For example, when a client is able to virtually explore a space, they may notice details that the architect didn't necessarily think were that important – after all, it's the client's home, and we all value certain aspects of our homes more than other aspects.

What I like about these VR experiences is they can be used to demonstrate both the exterior and interior of a property, and to demonstrate not just how a space will look, but also how it will feel to be in there (remember the audio of roaring fires and birdsong from earlier in the chapter?). I bet this would reduce a lot of concern and stress that clients feel when they're embarking on a major project.

XR in Construction

Moving on from pre-construction planning and design, we come to the construction process itself. It's easy to think of construction as a very traditional industry, not famed for its innovation – indeed, the oldest operating business in the world is Kongo Gumi, a Japanese construction company founded in the year 578 – but that would be a mistake. The industry is constantly evolving and adopting new technologies, such as using 3D printing techniques to "print" houses in a matter of hours, or using drones to map construction sites. The industry is also embracing XR technologies, particularly AR, which is proving to be a useful visualization tool.

Visualizing construction projects to improve accuracy

In construction, mistakes can cost companies dearly (both in terms of time and money). So it makes sense that the construction industry has begun adopting AR-based visualization tools as a way to improve accuracy on projects. Typically, this would involve wearing AR glasses or an AR headset, which would superimpose visuals onto what the construction worker is seeing in real life. For example, AR graphics could display exactly where a brick wall is supposed to go, and where the wiring will be located behind that wall, giving the tradesperson precise, real-time images and measurements to follow. Essentially, AR gives a more accurate view of what is being built.

This ties in with the rise of building information modeling (BIM), the process of using intelligent 3D models to plan and build projects more effectively. The idea is AR can enhance the use of BIM by turning complex models into easy-to-follow visuals, ensuring everything fits where it is supposed to. It is even possible to highlight hazards and danger zones in AR (I talk more about safety next).

More and more companies are combining BIM with AR to bring 3D plans to life on-site, using software such as Intellectsoft, which is compatible with Microsoft's HoloLens headset. (Proof that Microsoft sees the construction industry as a serious market for its product, the HoloLens is certified as basic protective eyewear, and you can even buy a HoloLens hard hat.) It is, however, possible to use AR without expensive hardware. UK-based engineering, environmental and building control consultancy MLM Group has used WakingApp's AR app (since acquired by Scope AR) to create 3D models of projects. Then, when the team is on site, they can view the models in MLM's own app and superimpose them on top of the original 2D blueprint.

Improving site safety

Despite strict safety measures and regulations, construction sites remain some of the most dangerous places to work in the world. (Here in the UK, construction is the third most dangerous industry to work in, and the rate of fatal injuries is around four times higher than the average across all industries.)[ii] Construction sites are dynamic places, constantly in flux, and each site is different, which makes identifying every possible risk and hazard near-on impossible.

When it comes to safety planning, research has shown that VR can help to improve safety by allowing professionals to visually assess conditions and recognize possible hazards before construction even starts.[iii] Crucially, these more immersive tools can simulate actual site conditions in a virtual environment, which makes them more effective than looking at standard 2D drawings. The same research also found that visualization technologies were very effective in immersive construction safety training and education, which makes a lot of sense. (Circle back to Chapter 6 for more examples from training and education.)

And when construction is under way, AR comes into its own as a safety visualization tool. Naturally, construction workers can't be immersed in VR headsets and blocking out everything around them on-site. That would be a pretty terrible idea. But AR glasses are transparent, so the worker stays aware of and engaged with their surroundings, while being fed additional data that's being superimposed over the real world. So, graphics can be overlaid that show where the wiring is, for example. Or, when a worker sees a real-life hazard sign somewhere on the site, AR glasses can display text to explain what the hazard is and what safety measures need to be taken.

Inspecting sites

Construction sites go through a rigorous round of various different inspections. Currently, these are done in a very analogue way – an inspector attends the site with a printout of plans or a checklist, manually inspects the site, takes photos and completes the required documentation. XR has the potential to enhance this process.

Research institute SRI International had a vision of developing a tool to make site inspections more efficient. Partnering with Japanese construction firm Obayashi Corporation, they created an AR system that captures the 3D information of built rebar and compares it to 3D building models, in order to identify any discrepancies between the built structure and the original plan (because discrepancies could potentially mean that the structural integrity of the building is undermined). With this system, a two-person manual inspection process has been transformed into a much more streamlined process that a single user can complete easily.[iv] For example, instead of taking photos of the rebar with large rulers held next to it (the standard process), the AR system automatically captures the diameter of rebar and compares the results against the plan. It's no wonder SRI International believes that AR has the potential to transform inspections, save time, cut costs and boost safety.

Lessons We Can Learn from Real Estate and Construction

It's relatively early days for XR in real estate and construction, but I hope this chapter has shown the enormous potential of XR to smooth out processes, cut costs and increase client satisfaction across these industries. Here are some lessons I think we can all learn from this chapter:

• Firstly, as the use cases in this chapter demonstrate, VR and AR can be used to achieve lots of different outcomes, for example, to aid the marketing of properties, enhance the building design process or improve safety on construction sites. When a technology can be used in multiple different ways, it's really important you deploy it strategically. Regardless of whether you're in the real estate or construction business, you really need to be clear on what you're aiming to achieve through XR. In other words, first you need to understand your goals, then you work out whether XR technologies can help you achieve those goals. This goes for any new technology, really, but it's worth emphasizing here.

• When you're dealing with people's property, you're obviously handling sensitive personal data (addresses, tenants' names, etc.). If your XR use involves the collection and processing of personal data, you will need to take great care to secure that data and ensure it isn't open to theft or misuse.

- I particularly liked the way VR can be used as a tool for collaboration, and I think this has potential beyond architecture and design. Consider whether VR could help your teams to collaborate across different locations, create together and provide context for new ideas.

Key Takeaways

In this chapter, we've learned:

- Across real estate, architecture and construction, VR and AR are helping to save time, reduce costs, enhance the client experience, and streamline processes. Although it's relatively early days for XR use (especially in construction), we can expect these technologies to play a greater role in future.

- In real estate, VR is allowing agents to offer immersive virtual tours that buyers can enjoy without having to leave home. Meanwhile, AR enables the virtual staging of properties and can guide tenants through routine queries and repairs.

- In architecture, VR is being enthusiastically adopted as a way to immerse clients in designs. Because clients can see and experience their property in VR before it is built, they have an opportunity to make changes earlier, which saves time and money further down the line.

- Finally, in construction, AR is proving useful as a visualization tool to improve accuracy, provide safety and hazard warnings, and make site inspections more efficient.

Just as architects are using VR to create immersive experiences for their clients (allowing them to "test drive" a building before it is built), hotels and travel operators are doing a similar thing in the travel industry. Turn to the next chapter to explore the exciting world of virtual trips and XR-enhanced travel.

Endnotes

i. Kingdom uses augmented reality to help with tenants' repairs; Scottish Housing News; https://www.scottishhousingnews.com/article/kingdom-uses-augmented-reality-to-help-with-tenants-repairs

ii. Workplace fatal injuries in Great Britain; Health and Safety Executive; https://www.hse.gov.uk/statistics/pdf/fatalinjuries.pdf

iii. Role of Visualization Technologies in Safety Planning and Management at Construction Jobsites; Procedia Engineering; https://www.sciencedirect.com/science/article/pii/S1877705817303399

iv. A modern approach to building inspections; SRI International; https://www.sri.com/case-studies/a-modern-approach-to-building-inspections-using-augmented-reality-and-mobile-technology-to-reduce-construction-overhead/

10
TRAVEL
AND HOSPITALITY

This is a sector that I'm passionate about and one that's suffered immensely in the wake of the coronavirus crisis. But there are reasons to be hopeful – one of which being the potential to improve and even transform the nature of travel and hospitality through VR and AR. As we saw in the education chapter, VR allows us to explore new places without physically traveling there, leading some to speculate whether virtual travel could ever replace the real thing. Personally, I don't think it will. But there's certainly value in being able to explore other parts of the world from home – not least because we can check out destinations before we travel there in real life. Virtual travel is therefore an exciting growth area to watch, but, as this chapter demonstrates, it's not the only application of XR within the travel industry.

Hotels are using VR as a marketing tool, by creating immersive virtual tours to tempt guests and show exactly what they can expect in real life (an idea that will appeal if you've ever arrived at a hotel only to discover it's not quite as wonderful as the pictures and description led you to believe). Plus, once guests have checked in, hotels are using AR to improve the guest experience through innovative, informative AR experiences (and even games). Then there are immersive navigation tools that help you find your way in a new place, thanks to super-clear AR arrows and directions that are overlaid onto the real-life street. And there are VR experiences that let you try out different excursions and attractions, so you can make the most of your limited vacation time. It's even possible to book your trip and hire a car using a VR platform.

XR technologies provide an opportunity to overcome some of the challenges and downsides of travel. Some suggest it could become an eco-friendly alternative to jetting around the world, thereby reducing the carbon footprint associated with travel. But we're a long way from virtual travel becoming the norm (if it ever does). At present, VR is great for providing an immersive snapshot (usually just a few minutes long) of, say, a hotel or a specific attraction, which the user can experience in beautiful 360-degree video. But it's not the same as being there in real life, feeling the breeze on your face, feeling the touch of an ancient stone wall, exploring the place in great detail, smelling the delicious smells wafting out of nearby eateries and sampling the local wine. That's why even the slickest VR experience can't replicate a culture-packed three-day minibreak or a relaxing two-week vacation on a beautiful island, and this chapter isn't going to pretend that it can. (That said, VR's ability to incorporate sensations like touch and smell is getting better, so we can hopefully look forward to some very exciting virtual trips in future.)

For me, VR really excels at bringing a place – be it a hotel, a city, an island resort or a specific landmark – to life. And this can be a powerful tool in the travel industry's toolbox, as many operators and hotels are already discovering. In other words, instead of deciding your next vacation destination and itinerary based on a booking website's description or by flipping through various travel guides, you can experience the place in a much more immersive way before you put your hard-earned money on the line. Try before you buy, basically. A VR-based sales pitch. And if you don't have a VR headset? No problem. Many of the VR experiences in this chapter are available as simple 360-degree videos that can be viewed through any web browser (although, obviously, the experience is much more immersive when viewed through a VR headset).

Then there's AR's ability to help us get more out of our vacation after we've arrived, using AR wayfinding tools and AR apps that give more information on local surroundings. As we'll see in this chapter, this creates enormous scope for travelers to personalize their trips and easily create their own self-guided tours – and steer clear of crowded, cookie-cutter city tours and excursions. For travelers, this is perhaps one of the biggest advantages XR can bring to the travel experience.

Therefore, the goal of this chapter is to showcase how VR and AR are helping the travel industry harness new business opportunities, ease customers' typical pain points, increase consumer confidence (and encourage bookings in the process), and generally improve the travel experience.

Virtual Travel

VR travel may not replace real-life travel anytime soon, but there are times when traveling to a destination is out of the question – whether you can't afford it or because a global pandemic has put the brakes on almost all leisure travel. Or sometimes you just want to check out different destinations before you decide on your next vacation. Virtual travel offers a handy travel taster, without having to pack a bag or update your vaccinations (until you're sure it's the right trip for you). Indeed, nowadays there is a wide range of travel experiences that offer exciting new perspectives for travel lovers.

Google Earth VR is a great starting point. Spanning the whole world, Google Earth VR lets you go anywhere you fancy. You can wander the streets of a busy city or fly through the air and enjoy a bird's-eye view of, say, the Hoover Dam or a mountain range. But Google Earth isn't the only option. Browse any VR app store and you'll find experiences to suit any type of traveler. For example, there's The Zion Narrows Experience, which transports you to the dramatic sandstone structures in Zion National Park, Utah. Or Rio 360, which lets you fly over Rio de Janeiro and discover the highlights of Brazil's most stunning city. Or Rome Reborn: The Pantheon, which explores the former Roman temple, including its design, decorations and the people who worshipped there. There are also numerous virtual tours of stunning islands in the Maldives – where my wife and I were lucky enough to honeymoon – from beautiful VR underwater experiences to virtual tours of hotels (more on that coming up later). Or there's Dubai360.com, produced by Dubai Tourism, which gives a bird's-eye view of the city, including famous landmarks such as the Burj Khalifa Pinnacle and the Palm Jumeirah.

But let's dwell on some VR experiences in a little more detail. Patagonia on Oculus Rift is one that provides a particularly rich experience. Exploring the mountain landmark of Monte Fitzroy, and specifically Laguna Sucia, the remote glacial lake that lies at the foot of the mountain, this VR experience gives users access to a stunning glacial lake that's hard to reach in real life. The experience combines 360-degree video with elements of gaming, so you can take off and soar around the turquoise lake, as if you were a bird, while listening to narration that explains the area's culture, history and geology. (Narrators Michael Breer and Luke Farrer – of Specterras Productions, creators of the experience – also describe the challenges of filming in such a tricky location.) And you can stop and explore the lake from a number of different vantage points, including the lake's shoreline and the ridge above it.

I also like The Redneck Road Trip, a 360-degree VR tour of the Badlands and backroads of California, created by award-winning landscape photographer Rod Edwards. Through this experience, users can embark on a road trip through this fascinating region of Southern California – which Edwards describes as a photographer's paradise – and live out their armchair wanderlust fantasies. As well as incredible quality visuals, there's a hand-illustrated map of the area, information on key locations and real audio recordings from locations. And because it's compatible with Google Cardboard and a mobile phone, you don't even need expensive hardware.

VR can also be used to raise people's awareness of real-world issues, which is the idea behind Greenpeace's VR video, Voyage to the Antarctic with Javier Bardem. This 360-degree VR video lets you accompany actor Javier Bardem as he joins a Greenpeace expedition to study the Antarctic. Although this isn't the first time Greenpeace has used VR video in their work – two previous VR videos include A Journey to the Arctic and Munduruku, featuring an indigenous community in the Amazon rainforest – the Antarctic video is particularly impressive because it includes a submarine dive to the bottom of the Antarctic Ocean, shot in VR. Greenpeace's goal with these videos is to use the immersive nature of VR to help people connect with environments more deeply and highlight the threats faced by delicate, at-risk habitats – basically by putting people "in" these environments.

An alternative way to look at it is VR provides a way for people to visit and engage with unspoiled destinations, without the damage that tourism can bring to pristine natural environments. VR may also provide a way for us to visit places after they have disappeared – gloomy as that thought is. The Maldives, for example, is facing a major threat in the form of rising sea levels, with some estimates predicting much of the nation could potentially be underwater within the next century. If we do lose precious habitats and natural wonders, from island nations to rainforests, VR could be our best way to visit and remember – and hopefully be driven to preserve remaining wonders.

Enjoying Virtual Tours of Hotels

One of the most common ways VR is used in travel is via "try before you buy" experiences. And not just from a destination perspective. High-end hotels are also using VR to show off their facilities and surroundings, let potential guests poke around rooms and generally tempt tourists to hit that "book now" button.

Among the adopters of virtual tours include a number of resorts in the Maldives. (Many of these tours can be experienced using just a smartphone, tablet, or computer, meaning you don't need a headset to explore your next holiday destination.) One such hotel is the Anantara Dhigu Maldives Resort, which lets you explore the different rooms (from beach villa to an over-the-water oasis), experience the many water sports on offer (surfing, sailing and jet-skiing), and check out the island itself. Or there's Furaveri Island Resort & Spa, where you can not only peek into rooms, you can also explore the island's stunning coral reef, near Hanifaru Bay, which is a UNESCO biosphere reserve.

Atlantis, the Palm in Dubai is another upscale hotel that boasts a stunning 360-degree panoramic VR video. Situated on The Palm Jumeirah, the famous man-made palm-shaped archipelago, this five-star hotel has a private beach, an underwater aquarium and even underwater suites that let you get up close to the marine life from your bedroom and bathroom window. The VR video provides a whistle-stop tour of the hotel's key features, taking in the impressive lobby, the Royal Bridge (the biggest suite in the hotel), one of the underwater suites, the famous Nobu restaurant, the aquarium, pool, and waterpark, and finishing with a nighttime stroll around the gardens.

Obviously, going to the Maldives or staying in a five-star hotel in Dubai is a pretty expensive vacation, so you want to make sure you're going to enjoy the experience. (And not turn up to find out your "luxury beach bungalow" is a ramshackle shed on a polluted beach, or that the hotel is half-built.) What I like about these virtual tours is that they inspire consumer confidence and encourage bookings. Which brings me onto the next use of VR in travel . . .

Booking Travel in VR

As well as encouraging customers to try before they buy, some operators are now encouraging bookings by allowing customers to actually book their trip through VR, as opposed to clicking on a mouse or screen.

Amadeus IT Group created Navitaire, the world's first virtual reality travel search and booking platform. The platform allows users to spin around the globe and visit different destinations. Then when they've decided where they want to go, they can search for flights, choose their seat on the plane, and book and pay for their trip – all using VR. For example, you touch dates on a virtual calendar instead of typing in your travel dates, and tap a payment device with a virtual version of your credit card to pay. The platform even lets you try out and book different hire cars. Watch a YouTube clip of the platform in action and it may seem a little clunky for now, but it's easy to see how this sort of technology could seamlessly integrate with "try before you buy" travel experiences. In other words, you could take a tour of a hotel, compare different types of room, check out the local attractions, and then book your stay and flight, without leaving the VR experience.

Test Drive Trips and Excursions in VR

As well as checking out different hotels and even booking your flights in VR, one of the best travel-related ways to use VR is to test drive different trips and excursions on offer in your chosen destinations. In other words, you can try out various experiences to decide how you really want to spend your time and money when you arrive on holiday. And these days, there are VR versions of many different excursions and sights around the world, such as the Pantheon VR experience I've already mentioned.

Or, you may find you have some people in your party who want to do certain excursions, while others don't. (Anyone who frequently goes on multi-generational family vacations, as I do, will know what I'm talking about here.) Diving, for example, doesn't appeal to everyone. Together, you could all have a go at a virtual diving experience, which may win over those who were otherwise unsure about delving below the water's surface. And even if it doesn't persuade them to join the real dive trip, they can enjoy the virtual version, you can enjoy the virtual and real versions, and you'll all have enjoyed some sort of shared vacation experience. At the very least, they'll know what you're enthusing about when you return from your real dive!

While I'm on the subject, there's a very cool virtual dive experience to a ship which sank in 1659 off the tiny Icelandic island of Flatey. The wreck, discovered in 1992 and Iceland's oldest shipwreck, is of the Dutch merchant ship Melckmeyt (Milkmaid), which sank during a storm in 1659. Now, anyone can take a virtual dive to the shipwreck and get a 360-degree view, with areas of the ship clearly labeled and explained. The virtual dive was created as part of a collaboration between maritime archaeologists and museums in Reykjavik, Iceland. Assuming diving in chilly Icelandic waters isn't your idea of vacation fun, you can enjoy the three-minute virtual dive by visiting the Reykjavik Maritime Museum and putting on a VR headset (there's a YouTube version available as well).

Virtual excursions of all kinds can help holidaymakers get more out of their vacations. But VR trips can also encourage bookings in the first place – for example, a customer may be more inclined to book a vacation once they've got a better idea of the fantastic excursions on offer in a resort, whether it's world-class diving or incredible cultural sights.

In fact, this is an idea British travel agent Thomas Cook was experimenting with back in 2015. Their "Try Before You Fly" VR experience allowed in-store customers to test-run different holiday experiences via immersive 360 VR films – with the goal of getting more customers to book then and there. Available at flagship Thomas Cook stores in the UK, Germany and Belgium, customers were invited to put on a Samsung Gear VR headset (Samsung partnered on the campaign) and experience a range of different destinations, including Egypt, Greece and New York. Some of the VR experiences included a helicopter tour of Manhattan and a trip to the Egyptian pyramids. The campaign was so successful, Thomas Cook reported a 190 percent uplift in bookings for New York vacations after customers tried the firm's five-minute New York VR experience,[i] demonstrating how VR can provide a significant boost to bookings.

Finding Your Way When You Get There – Immersive Navigation

Once you arrive at your destination, you can now use AR to help you find your way around. The idea is simple: using AR to overlay navigational instructions onto the real-life street or location that's in front of the user. Simple, yes, but it's small advances like this that help to make international travel a heck of a lot easier. So, if you ever end up lost in a busy foreign city, or if you simply struggle to make sense of 2D maps, these cool route-finding tools could help.

As the go-to navigational app for most of us, it's no surprise that Google Maps now incorporates an AR feature for those who are navigating on foot. Called Live View and announced in 2019, Live View is available on all ARCore and ARKit-enabled mobile devices, and in any locations where Google already has Street View. This new feature overlays big arrows and easy-to-follow directions onto the real world, to guide users on which direction to walk and where to turn. (At the time of writing, it's not available for users who are driving.)

Another example comes from Tunnel Vision NYC, an AR app that turns New York City's subway maps into interactive visualizations. Users point their phone at a New York Metropolitan Transportation Authority map and see information overlayed over the map, including info on transportation and local neighborhoods.

The airport experience can also be enhanced with AR. Particularly in big airports, such as London Gatwick, finding your way and getting to your gate on time can be a challenge, or at least can be a source of anxiety for nervous travelers. So Gatwick launched a passenger app to help. Harnessing more than 2,000 wayfinding beacons throughout the airport's two terminals, the app provides AR maps that help users navigate through the airport using their mobile phone. The app – which went on to win the Mobile Innovation of the Year and Mobile App of the Year awards[ii] – also keeps passengers up to date with personalized, real-time flight info and security queue times.

Similarly, Chinese ride-sharing app DiDi has an AR feature that guides passengers through busy buildings to find their exact pick-up location. I don't know about you, but I've regularly arrived at a huge airport or train station, with multiple exits, and not been able to find the taxi driver that's ready and waiting for me. DiDi's app-based AR navigation service solves that problem by guiding passengers through large buildings such as airports, train stations and malls to reach their driver.

Much as most of us love travel, and are incredibly fortunate to be able to experience different parts of the world, we'd be lying if we said travel was entirely devoid of annoying bugbears. As these tools show, AR can help to remove the minor annoyances that go along with travel and help travelers enjoy their trip even more.

Improving the Destination Experience

You've arrived on vacation, you've successfully navigated the airport, made it to your hotel, and have oriented yourself in your immediate surroundings. Now it's time to start taking in the local sights. So you point your phone at a landmark or attraction – whether you're stood in front of it in real life, looking at it on a map or looking at a photograph on your hotel room wall – and get helpful information on the attraction's location, history, opening times and entry fee. This is what AR brings to the travel experience, and I think it's an area that will grow and improve massively in the next few years.

One good example comes from City Guide Tour, an AR app that provides information on local points of interest, such as location, description, opening hours and admission price. At the time of writing, it has city guides available for Dubai, Prague and Torun, Poland, with more cities, such as Paris, coming soon. Or there's Tuscany+, the official AR app developed by Tuscany's tourism board. In live view, you simply click on icons to get details about a location, landmark, museum or even a restaurant, and points of interest are color-coded by dining, accommodation, sight-seeing and entertainment. My family and I used a similar AR travel guide when we were in Florence and, while there was definitely room for improvement, it was a useful tool. Essentially, AR apps like these allow travelers to conduct their own personalized, immersive, self-guided tours, without missing out on informative content. In the future, I think we'll see AR travel guides become much more mainstream, more detailed and more interactive.

Hotels are also using AR to improve the guest experience, offering information on demand, to make the guest's stay more enjoyable.

Evidence suggests technology like this is of growing importance to travelers – a survey by hotel management platform ALICE found that 43 percent of travelers want in-room technology that integrates with their personal devices.[iii] But what sort of technology are we talking about? As an example, British hotel chain Premier Inn, the UK's largest hotel brand, has incorporated AR into its Hub Hotels in London and Edinburgh; rooms feature an interactive, stylized map of the city – and when scanned with a smartphone, the map displays local attractions and information on points of interest. Elsewhere, the eco-friendly Olive Green hotel on the Greek island of Crete has its own app to enhance the guest experience. As well as being able to check in and out, control everything in their room, and access a "digital concierge," the app also allows guests to scan their QR-coded headboard (each room's headboard features an image of a beautiful Cretan landscape) and receive information on the destination and directions on how to get there.

It's not just about being informative, though. AR can also be used to deliver playful, fun experiences for guests. Marriott Hotels, for example, partnered with LIFEWTR by Pepsi (the official in-room bottled water in Marriott Hotels in the United States) to create an arty AR experience. After scanning the customized tags on the water bottles with their phone, guests could customize their rooms virtually with art from 18 different artists commissioned specially for the project. Alternatively, guests could create their own individual digital artwork for their room. And, like many of the AR experiences featured in this book, of course there was an option to share AR creations on social media.

Hotels can also create their own game-like experiences, inspired by the likes of Pokémon GO. Back in 2016, the Best Western Kelowna hotel in Canada worked with Canadian AR and software specialists QuestUpon to create an interactive wildlife adventure on the hotel's grounds. BC Wildlife Adventure Quest, as the game was called, used

AR to superimpose Canadian wildlife in the hotel's courtyard – including a moose, a grizzly bear and a creek with salmon jumping out of the water.

British travel and tourism group TUI has also been experimenting with using AR to enhance destination experiences in Palma de Mallorca. In the pilot project, which was conducted in 2019, travelers were kitted out with AR glasses that displayed information, video and images about local attractions so they could learn more while wandering around the beautiful city. According to TUI, AR will allow guests to explore places individually (so, no more crowded tours, following a guide with a bright umbrella), without missing out on insider knowledge – and I think this is a key selling point of AR in travel. I write a lot about business and technology trends and the move towards more personalized products and services is a major trend across most industries. It makes sense, then, that the travel industry is also seeking to provide more personalized experiences for guests, whether that is inside the hotel or when travelers are out exploring.

Looking at the wider hospitality sector, restaurants and bars are also getting in on the act. Remember the VR cocktail from Chapter 5? There's also an AR version: Mirage by City Social, an AR cocktail experience devised by London Michelin-starred restaurant, City Social. The experience paired real-life cocktails with a purpose-built AR app (which customers were prompted to download while waiting for a drink) and special coasters designed to trigger the AR visuals. So, when customers were presented with their drink, served on the special coaster, they had to open the app and point their phone at the drink – at which point they would see artistic designs digitally overlayed around the cocktail. Each AR cocktail on the menu represented a different artist (such as Banksy), so customers would see different animations depending on what drink they ordered – and, naturally, the designs could be captured as a photo or video, ready to share on social media.

In restaurants, AR can be used to project images of food onto your table. Think of it as the food equivalent of "try before you buy." In New York, restaurant chain Bareburger partnered with AR 3D model specialists QReal to create AR versions of Bareburger dishes, allowing customers to see a 3D model of a meal before they order it. According to QReal co-founder Alper Guler, the idea for the technology came about when he was struggling to explain Turkish dishes to friends. It's easy to see how this sort of technology could be really useful when you're traveling somewhere new, you've never experienced that culture's food before and have no idea what certain dishes are.

Meanwhile, other restaurants have used VR to create an incredible dining experience. One of the most famous examples comes from Sublimotion, a 12-course haute cuisine experience in Ibiza, Spain, that costs around $2,000 per person. Created by double–Michelin-starred chef Paco Roncero, the VR dining experience combines food with art, music and VR to create a three-hour extravaganza – and one of the most expensive meals in the world.

Clearly, VR and AR cocktails and meals are still something of a novelty, and may remain that way, but it goes to show how XR technology can be used to create a new kind of experience for travelers, diners and guests.

Lessons We Can Learn from Travel and Hospitality

I hope you've been inspired by these fascinating use cases from the world of travel and hospitality. But how can these examples add value in your organization? For me, the important lessons from this chapter include:

- As we've seen elsewhere in this book, "try before you buy" is one of the most powerful ways to use VR and AR. Whatever sector your business operates in, consider whether your customers would welcome the ability to immerse themselves in your product, service or experience before they buy.

- The AR wayfinding tools in this chapter really highlight how XR technology can remove the annoying wrinkles from everyday life. Think carefully whether VR or AR could help to overcome your customers' biggest pain points or challenges.

- I also love how XR is enabling the travel industry to create more personalized experiences for travelers. This is a huge trend across many industries. Could XR tools allow your business to provide a more personalized offering, or allow your customers to create their own unique experience?

- In general, what I've learned from this chapter is that, despite the huge challenges faced by the travel industry in recent years, many are embracing VR and AR as a way to stay ahead of the competition and differentiate their offering. This is particularly true of high-end resorts and hotels, like those in Dubai and the Maldives. So if your business positions itself as a high-quality provider, XR could be a way to emphasize that message and create a competitive advantage.

Key Takeaways

In this chapter, we've learned:

- While virtual travel may never replace real travel, VR and AR are certainly helping to improve the travel experience.

- Virtual hotel tours and VR excursions show the traveler exactly what they can expect from a hotel, destination and different attractions. For tourists, this inspires confidence (much more so than photos and descriptions, which can be deceiving). And for tour operators and hotel brands, this can be a valuable marketing tool that boosts bookings. In fact, it is even possible to book trips using VR platforms, so you can check out a destination and make your booking without having to leave VR.

- Meanwhile, AR can be used to create immersive navigation tools (such as Google Maps, featuring AR arrows and directions) that make finding your way around a new place easier than ever. Hotels are also using AR to enhance the in-room experience and provide guests with more information on local attractions. Plus, AR can be used in restaurants to show guests what dishes look like (in 3D) before they order.

Now let's turn to something very different and see how companies in industry and manufacturing are using XR technologies to drive business success.

Endnotes

i. Thomas Cook Virtual Reality Holiday "Try Before You Fly"; Visualise; https://visualise.com/case-study/thomas-cook-virtual-holiday

ii. Gatwick's Augmented Reality Passenger App Wins Awards; VR Focus; https://www.vrfocus.com/2018/05/gatwick-airportsaugmented-reality-passenger-app-wins-awards/

iii. Hotels' Digital Divide; ALICE; https://www.aliceplatform.com/hubfs/ALICE-Hotels-Digital-Divide.pdf

11

INDUSTRY AND MANUFACTURING

Perhaps some of the biggest opportunities for companies to leverage XR technologies lie in industrial and manufacturing settings. In fact, research by PwC indicates that in product and service development alone, the use of VR and AR could deliver a $360 billion boost to GDP by 2030.[i]

Speeding up time-to-market is one of the key uses identified by PwC research, and as we'll see in this chapter, many manufacturers are using VR and AR to improve their product design and development processes. Elsewhere in manufacturing, XR technologies can be used to plan production processes and facilities, train and inform assembly staff, speed up the production process (while reducing errors), and improve maintenance and inspection workflows. XR is also being leveraged in the oil and gas sectors, particularly to provide instructions to workers in remote or difficult-to-access locations (on a deep-water oil platform, for instance). And in logistics, companies are using AR to enhance the order picking process.

Depending on the setting, AR is proving particularly beneficial for manufacturing, industrial and logistics companies, because the user can stay aware of their surroundings and remain "in touch" with the real world. (You wouldn't want workers surrounded by heavy, dangerous and expensive equipment to be immersed in a VR headset.) This ties in with research by Capgemini, which found that 66 percent of organizations believe AR to be more applicable and relevant to their operations than VR,[ii] reflecting the fact that AR can be used to enhance interactions with machines in the real world, while VR generally seeks to isolate the user from the real world. The same research, unsurprisingly, found that implementation of AR was higher than VR.

If you tend to think of manufacturing and industry as slow to adopt new technology, think again. Industries like oil and gas have had no choice but to adopt emerging technologies in order to maintain pace with the greener energy solutions that are becoming available. And in manufacturing, the widespread use of robotics, automated production lines and artificial intelligence shows that the sector is no slouch when it comes to investing in tech. So, could XR be the next big technology trend in industry and manufacturing? If the use cases in this chapter are anything to go by, I think that's a reasonable claim.

Ultimately, I hope the examples in this chapter show the very real benefits XR is bringing to manufacturing and industrial organizations – indeed, many of the companies featured report significant cost and time savings after adopting VR or AR. Capgemini's research backs this up, noting that at least 75 percent of companies yielded more than 10 percent operational benefits in areas such as increased efficiency, productivity and safety. In other words, XR has gone way beyond hype and is delivering real value and competitive advantage. In this chapter, we explore what that means in practice.

Enhancing Product Design and Production Planning

Competition is tight in the manufacturing sector, and the ability to innovate and bring new products to market fast is key to success. This is where VR and AR can enhance the product design process, essentially by helping to speed up the creative process. Like the architecture examples we saw in Chapter 9 – where 3D models can be ported into VR – in manufacturing, product designs can be explored in virtual or augmented reality, reducing the need to build expensive and time-consuming physical pretotypes and prototypes. Ideas can be tested more quickly, in other words, to identify what works and what doesn't. Plus, as we saw in Chapter 9, VR can also enhance the collaborative process, giving creatives a virtual space to design and share ideas or feedback, regardless of where they are physically located. Let's look at how XR technologies are enhancing the design process.

Thyssenkrupp

German engineering giant Thyssenkrupp has used Microsoft's Holo-Lens headset to improve the design process for its highly customized home mobility solutions. Thyssenkrupp's traditional product design process used to involve multiple, complex stages of measurement (including cameras and manual data entry), so that the customized stair lifts could be precisely tailored to each customer's home. While effective, this resulted in lengthy wait times for customers. But with the HoloLens, a salesperson is able to measure the customer's staircase in one visit, and provide the customer with a digital visualization of what the stair lift would look like in their home. Then the intricate measurement data is automatically sent to the manufacturing team via Microsoft's Azure cloud platform, eliminating the need for manual data entry. This improved system reduced delivery times by as much as 400 percent.[iii]

XR in car design – Ford and Jaguar Land Rover

American multinational automaker Ford has also used the HoloLens to design cars in mixed reality. Using HoloLens, designers can quickly model changes to vehicles, viewing the changes on top of an existing, physical vehicle, which is much faster than the traditional process, which involves making clay models. While Ford still uses clay models in many cases, the HoloLens allows designers to quickly try out new ideas without having to make new clay models for every design.

I was also interested to see Ford was using VR to keep the design process moving during the coronavirus pandemic, when designers had no choice but to work from home. Using VR headsets, design leads were able to log in to a virtual studio and inspect progress on new designs, share ideas and generally work in a more collaborative way, despite the physical distance. Even before the pandemic, they were using VR to view new designs as a team (back when they were in the same room) – for example, viewing a vehicle in different settings and lighting, so they could make design adjustments before creating a clay model.

Using XR in vehicle design is nothing new. In fact, as far back as 2008, British automotive company Jaguar Land Rover was using an innovative VR engineering and design studio – dubbed "The Cave" because it looked like an empty room – to design cars. Because designers could visualize full-scale 3D models of new designs, inside and out, they were able to reduce the number of physical prototypes needed, which in turn saved time and money. So much so that "The Cave" saved Jaguar Land Rover more than three times its cost within just two years of operation.[iv]

Production planning

XR technologies can also benefit the production planning process, whether that's planning where to place personnel and equipment in a new factory setting or simply planning how a new product line will be built. For example, at American aerospace manufacturer Boeing, mechanics used VR to prepare for building the new 737 MAX 10. Using VR headsets, mechanics could see how the landing gear should be installed, and what kinds of tools would be needed, which gave them a chance to provide feedback on potential pinch points or new tooling equipment months before assembly actually started.

But the scale can be much grander, to the point where an entire factory floor is modeled in VR to check that everything is placed and connected in the best way. In this way, virtual plants or production lines can be used to test workflows before production changes or new products are implemented in the real world. This is critical because there's so much to get right in manufacturing – space needed for assembly lines, safety distances, equipment dimensions, etc. – and if you get it wrong, a delay or shutdown can be very costly.

AR can also be useful here, by overlaying visuals over an existing factory space. This is what German automaker Volkswagen did when they used AR to remodel production lines at its plant in Chattanooga, Tennessee. Using AR headsets, engineers could model how equipment would interact in the real-world environment of the production line, which helped the team spot pinch points between machinery that otherwise wouldn't have been easy to identify.

German multinational chemical company BASF also uses AR factory planning software to merge the digital and real worlds, and generally speed up planning processes. In other words, pipework and assembly equipment that doesn't exist yet can be visualized on-site, directly in the real-world setting. What's really cool is alterations to the design can be made in real time, allowing engineers to test out different "What if?" scenarios.

Training Workers Using VR and AR

I've already talked a lot about training and education in Chapter 6, but because training is a key use of XR in manufacturing and industry, I wanted to highlight two relevant examples here.

Honeywell

Multinational engineering, industrial and aerospace conglomerate Honeywell uses VR and AR to address the skills gaps – or, specifically, the problem of knowledge "leakage," where older workers retire and take their knowledge with them. Traditionally, retirees were asked to put their knowledge into PowerPoint slides or Word documents that could be shared in classroom-like spaces with new hires. But Honeywell found that this passive learning experience doesn't lead to great knowledge retention for new employees – after three months, information retention was around 20–30 percent. So they equipped both departing workers and new hires with the HoloLens mixed reality headset. Retirees could record exactly what they were doing in their work, and then new workers could see this information overlaid onto their own work activities. This more active form of training boosted the level of information retained from 30 percent (at best) to 80 percent.[v] Honeywell also uses the same technology to reduce maintenance costs for offshore platforms by up to 50 percent.

Rolls-Royce

Long-standing British engineering company Rolls-Royce has implemented immersive VR training for its business aviation customers. The remote training, which allows participants to service the Rolls-Royce BR725 aircraft engine, uses VR equipment (which is shipped to the customer's door) to immerse the customer in an instructor-led, two-day distance learning course. The immersive VR environment allows participants to get up close to a virtual version of the engine, interact with the engine and tools, watch the steps involved in a particular task and then virtually complete the task themselves, under the supervision of an instructor.

XR in Oil and Gas

Facing competition from clean energy solutions, the oil and gas sector has had no choice but to transform its processes and embrace new innovations. And XR technologies have a big role to play in this digital transformation, particularly when it comes to helping personnel conduct repairs in remote locations, such as offshore oil rigs. In other words, instead of transporting an expert to a rig, which takes time and costs money, AR glasses can give on-site workers detailed instructions on how to fix issues. Or, as we've seen elsewhere in this book, VR can provide workers with immersive safety training to prepare them for all kinds of scenarios – scenarios which would be too difficult, dangerous or costly to simulate in real life.

Royal Dutch Shell

One of the oil and gas "supermajors," Anglo-Dutch multinational Shell is using AR and VR in a number of ways. In one example, workers at the Malikai deep water oil production platform off Malaysia receive VR safety training. VR is being used to save time when new teams come onboard – because they can familiarize themselves with everything before they set foot on the platform – and generally improve the transfer of knowledge from one team to another. And in 2019, Shell announced plans to connect frontline field workers with back office expertise through the use of AR helmets, designed by RealWear. These look like regular hard hats, but with a micro-display and camera that sits a short distance from the wearer's right eye – so, their regular vision isn't impaired, but they can share and send pictures and video with office personnel, and get help with remote operations in real time. Shell plans to roll out the helmets in 24 operational sites around the world.[vi]

BP

Turning to another energy "supermajor," BP uses the HoloLens MR headset to improve its upstream operations (basically, exploring, finding and producing oil and gas). The headsets are used to give teams working in the upstream environment access to digital 3D models of outcrops. (Areas of the earth's surface that have experienced tectonic activity – there's a lot of jargon in oil and gas exploration!) So, in a very basic sense, drones can capture images of an area, and this data can be transformed into 3D models, which can then be visualized by teams in the area using the HoloLens headset, in order to inform the exploration strategy.

Although it may seem like the oil and gas industry deals with rather unique challenges, these examples show how VR and AR can be used to train and assist anyone working in dangerous or remote settings, regardless of the industry.

Deploying XR Across Production and Manufacturing Processes

When it comes to the production or assembly process, AR and MR can bring huge benefits. (VR not so much, since it's generally not a good idea to have people on an assembly line wearing VR headsets!) AR glasses like Google Glass or AR/MR headsets like the HoloLens can overlay instructions and graphics onto real-life components and products, thereby helping technicians and operators get up-to-speed quicker. Here's how it works in practice.

NASA and Lockheed Martin

American aerospace and defense company Lockheed Martin is the contractor charged with building Orion, the NASA spacecraft that will take humans deep into space. (As an aside, Lockheed Martin was a fairly early adopter of AR glasses, using them as far back as 2015 to build the F-35 combat aircraft. At the time, the company said the technology enabled engineers to work 30 percent faster with 96 percent accuracy.)[vii]

Now, Lockheed Martin engineers are using HoloLens headsets to build Orion faster, without having to rely on the thousand-page instruction manuals that are common in the aerospace business. The headset overlays holograms of models created in design software, along with labels and instructions for specific parts, onto the aircraft as it is being built.

Lockheed Martin found the headsets dramatically reduce the amount of time it takes technicians to get familiar with and prepare for new assembly tasks. And the company is even hoping that the technology could one day be used in space, to help astronauts maintain the spacecraft Lockheed is building.[viii] What's really interesting is Lockheed engineers are using the HoloLens on a daily basis. So how does it shape up for everyday use? Technicians report the longest they can wear the headset without it getting too cumbersome is about three hours, indicating that (for now at least) the technology is best suited to learning a particular task, solving a particular problem or checking on-screen directions periodically, rather than wearing all day. That said, the technology will get lighter and more comfortable in time. (Check out Chapter 13 for a more detailed look at the future of XR.)

BMW

Elsewhere, technicians in BMW service centers are using AR glasses to connect with engineers and other experts to solve complex repair and maintenance issues. Because the glasses provide a hands-free video link, the technician and expert can work through the issue together, solve the problem more efficiently and get customers back in their cars faster.

Tesla

On the subject of automobile companies, Tesla is one that's known for embracing new technologies (from electric vehicles to automated manufacturing processes). The company was using Google Glass in its factory as early as 2016, but in 2018 Tesla filed its own patent for Google Glass–type AR glasses.[ix] The idea is Tesla's glasses, which double as safety glasses, will assist those working on the production line by helping them identify places for joints, spot welds and interfaces with other parts. It'll be interesting to see whether this patent comes to fruition.

GE

American conglomerate General Electric spans the power, renewable energy, aviation and healthcare industries. At its Pensacola, Florida, factory, workers assembling GE wind turbines wear AR glasses that display digital instructions on how to install parts correctly – instead of having to stop and check manuals at regular intervals. According to GE, this delivered a 34 percent improvement in productivity versus operating in the standard way.[x] Through the glasses, technicians can also access training videos or, using voice commands, contact experts for further assistance (the workers can live-stream their point of view so the expert can see what they see). For me, this shows how

AR glasses are incredibly versatile, and can do more than just display instructions.

Boeing

Earlier in the chapter, I mentioned how Boeing used VR to prepare for manufacturing a new airplane. But that's not the only way the company uses XR. It has also tested AR as a way to give technicians hands-free access to interactive 3D wiring diagrams. As you can imagine, installing the electrical wiring on a plane is a pretty complex, high-stakes task, but AR glasses have the potential to make it easier. The traditional method of wiring a plane involves looking at 2D, 20-foot-long wiring schematics and interpreting those diagrams into the 3D aircraft in front of the technician. But with AR, technicians can move around the aircraft and easily see where the wiring is intended to go, as renderings are overlayed over the real-life fuselage. Boeing's studies showed the technology delivered a 90 percent improvement in first-time quality – and a 30 percent reduction in the amount of time spent doing the job.[xi]

Improving Quality Control and Inspections

That brings us neatly on to the use of XR in quality control and safety/ maintenance inspections. In Chapter 9, we saw how VR and AR can be used in virtual property tours and building site inspections, so the leap to manufacturing and industry inspections seems a logical next step for the technology. Let's see how companies are streamlining and improving the accuracy of inspections with the help of XR.

AGCO Corporation

American agricultural machinery manufacturer AGCO is using AR in the manufacturing and inspection of its tractors and other agricultural equipment. After a successful pilot study back in 2014, the technology has been rolled out more widely at the firm's manufacturing site in Jackson, Minnesota. Because AGCO manufactures equipment to customers' exact specifications, each machine is unique – creating hundreds of potential variations on top of the firm's one hundred basic models. As you can imagine, this level of variation has the potential to create an assembly and quality control nightmare! But using Google Glass AR glasses, assembly workers and quality inspectors can access manuals and instructions right in front of them, leading to a 25 percent reduction in production time on low-volume, complex assemblies and a 30 percent reduction in inspection time.[xii]

Airbus

European aerospace company Airbus – which builds military air lifters as well as commercial airliners – is overhauling its maintenance inspections process for military aircraft through the use of drones and AR. Drones fitted with high-definition cameras and AR LIDAR remote sensor technology conduct a fly-around inspection. The data generated can then be displayed on tablets and AR glasses, which allows experts to quickly identify any defects (artificial intelligence is also used to help spot defects) – meanwhile, the system also formally records all the inspection and maintenance procedures.

Traditional inspection methods involve building scaffolding around the aircraft so it can be inspected up close, which can cause inadvertent damage, so Airbus says this method creates a more robust maintenance process with lower risk of aircraft damage. And, impressively, the new system saves a lot of time, cutting the time for an external aircraft inspection from weeks down to just two hours.[xiii]

(As an aside, AR has also been trialed in other military settings, helping mechanics carry out repair work on military vehicles. Researchers at Columbia University have previously worked with mechanics from the US Marine Corps to develop an AR system that helped users carry out repairs on armored vehicles in half the usual time.[xiv] Check out Chapter 12 for more military use cases like this.)

Returning to Airbus, as well as leveraging AR, the company uses VR at the design and development stage to determine the best way to maintain an aircraft. Thanks to VR simulations, teams can check the feasibility of maintenance activities when the aircraft is being designed, allowing for modifications that increase reliability while minimizing maintenance costs. This process would traditionally be done with a computer-based modeling system and a digital mock-up, and later in the development phase through physical verification of various activities (for example, checking whether mechanics can access or remove a certain component). Now, using portable VR masks, that verification and validation can be done in just 25 percent of the time needed for traditional methods.[xv]

The Use of XR in Logistics and Warehousing

XR technologies – particularly AR glasses – can bring big benefits to logistics and warehousing functions. The most common uses here relate to warehouse planning (much like the factory floor planning examples mentioned earlier in the chapter) and order picking (guiding workers around the warehouse facility to ensure quicker, more accurate picking). Ultimately, the technology is being used to simplify and streamline warehouse processes, thereby boosting the supply chain as a whole.

DHL

DHL – or more specifically, its parent company Deutsche Post DHL – is the world's largest logistics company. The company successfully deployed AR for order picking in the Netherlands back in 2015 and is now investing in AR technology across North America.

Partnering with DHL customer Ricoh and wearable technology experts Ubimax, DHL carried out a pilot project in 2015 to test AR glasses in a warehouse in the Netherlands. The technology was used for "vision picking," meaning staff were guided through the warehouse by graphics displayed on the glasses (a somewhat fancier version of the AR wayfinding examples we saw in Chapter 10). Thanks to this technology, staff were able to speed up the picking process and reduce errors – so much so that DHL reported a 25 percent increase in picking efficiency.[xvi] Then in 2018, DHL announced plans to roll out emerging technologies – including AR – in 350 of its 430 facilities in North America. The huge investment was reported to be worth $300 million.[xvii]

Although the warehousing sector has, by and large, seen some great leaps forward in terms of emerging technology (such as robotics and automated picking systems), the truth is many warehouses around the world still rely on the traditional pick-by-paper approach, which is not only slow and prone to errors, it also relies heavily on personnel knowledge (not ideal when many warehouses depend on temporary workers). AR glasses can display a digital picking list (which means the worker can operate hands-free), guide the picker to the next item using the most efficient route and, thanks to barcode scanning or image recognition software, help them locate the correct item on the shelf. Even a small improvement in the picking process can deliver tangible benefits to a business, so this use case is particularly promising for any business who operates a warehouse facility.

Combining XR with digital twins

A digital twin is an exact digital replica of something in the physical world – anything from a warehouse to a small individual component, even a business system or process. Digital twins are created by connected Internet of Things (IOT) sensors that gather data from the real world and send it to machines to reconstruct digitally. The idea is that a company can use the digital twin to test different scenarios with much less risk, uncover insights about how to improve operations, or even spot potential problems before they occur in the real world (useful in a manufacturing setting). In essence, lessons learned from the digital twin can be applied to the real-world version to increase efficiencies while generally reducing risk and boosting return on investment.

The concept of digital twins has been around for a while now, but it's the explosion in smart, IOT devices and sensors that made digital twins a much more affordable and accessible option for businesses. And when information from these digital twins is visualized using AR or VR, the outcomes can be even more impressive.

For example, using AR, data can be fed to on-site workers in real time to warn them of a potential issue, overlaying the information onto the real world via AR glasses, or even a phone or tablet. Or by combining digital twins with VR, you can create immersive 3D representations of specific sites or scenarios. This is very much an emerging field, but I believe we will see greater integration of digital twin and XR technologies in the future, as companies look to leverage new technologies to deliver greater efficiencies.

Lessons We Can Learn from Industry and Manufacturing

If you're considering implementing XR in an industrial, manufacturing or warehousing setting, I believe the use cases in this chapter offer some valuable learning points:

- Remember, XR gives you the opportunity to plan and test in a more effective, immersive way – whether that involves designing or adapting products, simulating new workflows in a virtual space (as in the Airbus maintenance simulations), or simulating an entire factory floor.

- AR offers particular value in manufacturing, maintenance and logistics settings because it allows the workers to stay aware of their surroundings (as opposed to VR, which is designed to fully immerse the user in a digital environment).

- It was also good to be reminded that AR isn't just about overlaying graphics and instructions over the real world – using voice commands, smart AR glasses allow workers to connect with experts in another location, share what they're seeing in front of them in real time, and receive audio or visual instructions and guidance on what to do next. This is worth bearing in mind for any remote workers, or those who work in dangerous or difficult-to-access locations.

- That said, there are some limitations to consider. For one thing, most current AR and MR headsets aren't that comfortable for long-term use. This will change as the equipment becomes smaller and lighter but, for now, it seems that the most appropriate use is for periodically checking instructions during a task, following a set of instructions in real time or learning a task before undertaking it, as opposed to wearing the headset for a whole shift. (There will also need to be advances in battery life if devices are intended to be worn for more than a few hours at a time.)

- Some of the use cases in this chapter have obviously required huge financial investment, which will be a major consideration for most businesses. As with any new investment, you have to weigh up the costs of implementing the emerging technology versus *not* doing it (i.e., the future cost savings and efficiencies you might miss out on if you don't invest).

Key Takeaways

In this chapter, we've learned:

- Particularly (although not exclusively) in the automotive industry, VR and AR are greatly improving the product design process, helping to speed up the design and approval process, and allowing creatives to test out new ideas without investing in expensive prototypes.

- Training is another key usage of VR and AR in industry and manufacturing, with companies reporting the technologies aid the transfer of knowledge between teams, reduce knowledge "leakage" when older workers retire, and boost knowledge retention among new hires.

- Oil and gas "supermajors" have also been using VR and AR in training, as well as making remote operations more efficient by providing instructions to workers on rigs and in upstream locations.

- Elsewhere, in production and manufacturing, AR is helping to speed up assembly times and reduce errors by giving workers the instructions they need, without having to refer to paper manuals. Similarly, the technology is also helping to streamline quality and safety inspections, and maintenance/repair processes.

- Finally, warehousing and logistics functions are also benefiting from AR, particularly through the application of "vision picking," where picking staff are guided through the warehouse by AR glasses and led directly to the correct items. A study by DHL found the technology delivered significant improvements in picking efficiency.

A couple of times in this chapter – see the Airbus and Marine Corps examples – I've pointed to military applications of XR. Now let's delve into this sector in more detail, exploring how VR and AR are being used to enhance military (and law enforcement) operations.

Endnotes

i. How VR and AR are transforming manufacturing; PwC; https://www.pwc.co.uk/industries/manufacturing/insights/how-vr-and-ar-transform-manufacturing.html

ii. Augmented and Virtual Reality in Operations; Capgemini; https://www.capgemini.com/wp-content/uploads/2018/09/AR-VR-in-Operations.pdf

iii. Thyssenkrupp transforms the delivery of home mobility solutions with Microsoft HoloLens; Microsoft; https://blogs.windows.com/devices/2017/04/24/thyssenkrupp-transforms-the-delivery-of-home-mobility-solutions-with-microsoft-hololens/

iv. Jaguar Land Rover's Virtual Cave; Automotive Council UK; https://www.automotivecouncil.co.uk/2010/11/jaguar-land-rovers-virtual-cave/

v. The Amazing Ways Honeywell Is Using Virtual and Augmented Reality to Transfer Skills to Millennials; Forbes; https://www.forbes.com/sites/bernardmarr/2018/03/07/the-amazing-ways-honeywell-is-using-virtual-and-augmented-reality-to-transfer-skills-to-millennials/

vi. Shell Revamps Remote Operations with Augmented Reality Headset; AREA; https://thearea.org/ar-news/shell-revamps-remote-operations-with-augmented-reality-helmet/

vii. Lockheed Is Using These Augmented Reality Glasses to Build Fighter Jets; Popular Mechanics; https://www.popularmechanics.com/flight/a13967/lockheed-martin-augmented-reality-f-35/

viii. NASA is using HoloLens AR headsets to build its new spacecraft faster; MIT Technology Review; https://www.technologyreview.com/2018/10/09/103962/nasa-is-using-hololens-ar-headsets-to-build-its-new-spacecraft-faster/

ix. Tesla wants its factory workers to wear futuristic augmented reality glasses on the assembly line; Business Insider; https://www.businessinsider.com/tesla-patent-reveals-augmented-reality-glasses-for-factory-workers-2018-12?r=US&IR=T

x. Looking Smart: Augmented Reality Is Seeing Real Results in Industry; GE; https://www.ge.com/news/reports/looking-smart-augmented-reality-seeing-real-results-industry-today

xi. Boeing Tests Augmented Reality in the Factory; Boeing; https://www.boeing.com/features/2018/01/augmented-reality-01-18.page

xii. 2017 Assembly Plant of the Year: AGCO Leads the Field with Lean Technology; Assembly Magazine; https://www.assemblymag.com/articles/93996-assembly-plant-of-the-year-agco-leads-the-field-with-lean-technology

xiii. Airbus innovation for military aircraft inspection and maintenance; Airbus; https://www.airbus.com/newsroom/news/en/2019/05/airbus-innovation-for-military-aircraft-inspection-and-maintenance.html

xiv. Faster Maintenance with Augmented Reality; MIT Technology Review; https://www.technologyreview.com/2009/10/26/208625/faster-maintenance-with-augmented-reality-2/

xv. Stepping into the virtual world to enhance aircraft maintenance; Airbus; https://www.airbus.com/newsroom/stories/stepping-into-the-virtual-world-to-enhance-aircraft-maintenance-.html

xvi. DHL successfully tests Augmented Reality application in warehouse; DHL; https://www.dhl.com/en/press/releases/releases_2015/logistics/dhl_successfully_tests_augmented_reality_application_in_warehouse.html

xvii. DHL Supply Chain Invests $300M to Accelerate Integration of Emerging Technologies into North American Facilities; Deutsche Post DHL Group; https://www.dpdhl.com/en/media-relations/press-releases/2018/dhl-supply-chain-invests-to-accelerate-integration-of-emerging-technologies.html

12
LAW ENFORCEMENT AND THE MILITARY

Even if you aren't in the military or law enforcement, many of the applications in this chapter are really inspiring – demonstrating the sheer awesome capabilities of XR and how it can be used to improve decision making and keep people safe. I'm certainly excited by some of the innovations being used in practice, and those being trialed for future use.

For the most part, the operational examples in this chapter center on AR glasses or headsets, which can be used to visualize critical data and aid decision making, particularly when the user is under pressure, without distracting them from what's going on around them. (VR isn't overlooked in this chapter, though; I include use cases that show it is a valuable training resource for law enforcement and military services.)

How can AR benefit these sectors? In law enforcement, the FBI says that AR's ability to overlay information or images onto a person's real-world view can help officers accomplish a variety of tasks and assignments more efficiently – potentially to the point where one officer equipped with AR technology could complete the same work as three unequipped officers.[i] We'll cover real-world policing use cases in this chapter, but some of the AR applications mooted by the FBI include:

- Real-time language translation or intelligence about crimes and criminals for officers on patrol

- Real-time supervision of officers on patrol (remember, AR can be used to share a live feed with others remotely)

- 3D mapping of cities, building floor plans, sewer systems and public transportation routes to improve situational awareness

- Improved situational awareness during SWAT operations, including advanced optics, and thermal and infrared imaging of fleeing criminals

- Identification of friends or foes to eliminate friendly fire casualties

- Enhanced ability to gather information at crime scenes for criminal investigations

Bottom line: AR can help police solve and combat crime in new ways – which is essential given that we live in a time of rapid change and technological advancement. In fact, the FBI warns that AR could provide criminals and terrorists with new opportunities to disrupt society, making it even more essential that law enforcement agencies get to grips with this technology. It's relatively early days but, as the examples in this chapter show, this transition to AR-enhanced intelligence in law enforcement is under way.

Just as in society at large, warfare is constantly evolving and adapting in line with new technologies. So it's perhaps no surprise that, in the military – which is so often at the cutting edge of technological innovation – the AR market is estimated to reach almost $1.8 billion by the end of 2025 (up from $511 million in 2017).[ii] AR is useful in both training and combat environments (VR also has a role to play in training, but not in active combat), with some of the most common uses including:

- Tactical AR for improved situational awareness – helping military personnel identify their own position more precisely, locate others around them, and identify who is a friend and who is a foe

- Improved night vision and thermal imaging

- Improved data on a target, including how far away it is

- The ability to fire a weapon accurately without looking at the target, meaning troops can maintain cover while engaging the enemy

- Improved maintenance functions (turn back to Chapter 11 for more on this)

What's great about AR in tactical settings is the information is displayed within the person's line of sight, meaning they never need to look down at a separate device or lose focus on what's going on around them. In a dangerous situation, that can mean the difference between life and death.

Let's explore what all this means in real-life applications, starting with the police.

XR in Law Enforcement

AR's ability to overlay information on what an officer is seeing – whether that's on a busy street or at a crime scene – can bring many benefits in law enforcement.

Identifying suspects in real time

Particularly when combined with artificial intelligence (AI) and facial recognition software, AR smart glasses can be used to help police officers on the streets identify suspects. Sound far-fetched? Chinese AR specialists Xloong developed smart AR glasses for Chinese police back in 2017, and the glasses have since been adopted by Chinese law enforcement authorities at highway inspection points and airports, as well as in six local public security bureaus, including Beijing.[iii] Using the glasses – which look like sunglasses – police officers can access national database information such as facial recognition and ID card data and vehicle plate information, all in real time. The idea is to catch suspects and people traveling under false identities. Of course, there are huge privacy concerns around this sort of usage, and the technology is clearly open to misuse (identifying journalists or activists, for example). But there's no denying it is a powerful example of how AR can equip officers with valuable real-time information.

It's not just about identifying suspects, though. AR glasses can also be used to identify people who may be unwell. During the coronavirus outbreak, Dubai transportation police began using AR technology – smart glasses combined with thermal cameras, infrared rays and AI – to measure the temperature of passengers in transport stations. A similar system was in place at Rome-Fiumicino Airport in Italy, except this time the technology was incorporated into a helmet, rather than glasses. Similar to a police riot helmet, but with a tinted visor and a large external thermal scanning camera that points forward, the helmet can screen multiple people at once. When a passenger registers an elevated temperature, the wearer of the helmet is alerted with sonic and visual alarms. With parts of Italy ravaged by the first wave of COVID-19, being able to accurately detect people with elevated body temperature in Italy's busiest airport was crucial, not least because it helped to prevent infectious people from boarding planes. If we're unlucky enough to experience another global pandemic anytime soon, I expect wearable technology like this will play a much more widespread role in identifying infections and keeping businesses and public spaces open.

Enhancing criminal investigations

Given the privacy concerns, we may not see everyday police officers on the streets equipped with ID-scanning AR glasses – at least, probably not in Western countries. But one area where I think AR has huge potential is in criminal investigations.

Processing and preserving all the evidence at a crime scene is a slow, intricate process – and the first officer on the scene may not be the most qualified person to identify and preserve evidence. Mistakes in preserving physical evidence or securing a crime scene can hamper an investigation or, at worst, lead to the perpetrator escaping a sentence. This is where AR glasses or headsets can help. The tuServ mobile policing app, which works with Microsoft HoloLens and mobile devices, is designed to help police officers on the scene. The app maps out the crime zone, captures digital evidence and allows officers to place virtual markers, without disturbing the physical scene and potentially tainting evidence. The digital version of the crime scene can be shared with other detectives, so they don't have to be physically present at the scene, and can be used to (digitally) transport investigators back to the scene, so they can recall every detail of the scene long after it has been cleared.

Elsewhere, Dutch police have trialed using AR to help officers at the scene of an emergency, using either a smartphone or HoloLens to assess the condition of the scene. The system allows remote teams – who can view the scene thanks to a camera on the officer's vest – to virtually point the officer to specific elements (using arrows and notes displayed on the officer's AR device), and give instructions on what needs to be bagged or preserved for investigators.

Using XR to Train Law Enforcement and Military Personnel

As we've already covered training and education in Chapter 6, I won't dwell on too many training examples from law enforcement and military. But suffice to say this is a huge area of opportunity, particularly when it comes to VR training. From combat training to flight simulators to police response training, VR can be used to simulate a huge variety of scenarios in a realistic and safe environment.

Police training

One example comes from the VirTra VR police training tool that I mentioned in Chapter 6, which can be used to train police officers in hundreds of simulated scenarios. Crucially, interactions within these scenarios can be escalated or de-escalated to help learners practice decision making under stress, and learn when to appropriately use force.

Elsewhere, simulations can help officers learn to spot signs of domestic violence. Gwent Police in Wales have used VR to prepare officers for domestic violence callouts, allowing them to test out decision-making skills and identify controlling behavior.

There are also SWAT training tools, such as Apex Officer's VR tactical training simulator. Simulators like this allow learners to make mistakes that could otherwise be fatal in the real world, and learn from those mistakes in a safe way. What's more, simulations can be individually tailored to meet different team's needs, or even to design training around an individual officer's specific strengths and weaknesses.

Military training

One of the obvious applications of VR in military training is to make flight simulators more realistic. In fact, the US Air Force is turning to

VR to help speed up the process of becoming a pilot and address pilot shortages as experienced older pilots are tempted away to commercial airlines. As of 2019, the US Air Force had a shortfall of 800 active-duty pilots and more than 1,000 reserve pilots, and with that shortfall predicted to grow,[iv] the Air Force desperately needs to make its pilot training more efficient. One initiative to address this is the experimental Pilot Training Next (PTN) program, which incorporates VR and AR technologies to decrease the time and cost of training. The UK's Royal Air Force is participating in the program.

Marine Corps recruiters have also been exploring the use of VR – this time to attract potential pilots. In 2020, the Marine Corps Recruiting Command announced it wanted to buy six state-of-the-art VR flight simulators to use at recruitment events.[v] The advantage of these VR simulators over traditional flight simulators is that the VR units would be standalone and easy to transport to different events, whereas the traditional simulators have to be transported via a dedicated 35-foot truck!

But it's not just VR that can be used for training. The US Navy has tested an AR platform called TRACER (or, to give it its full name, Tactically Reconfigurable Artificial Combat Enhanced Reality) at the Center for Security Forces in North Carolina. Built largely using off-the-shelf gaming gear, including the Magic Leap One AR headset and a simulated weapon that delivers realistic recoil, the system allows trainers to create different training scenarios that can be overlaid onto real-life settings, and enables them to easily change the opposition forces and threat posed. Navy officials have also said the system will help them deliver better training on ships, where space is limited.[vi]

Operational Uses of XR in the Military

While VR may be very useful for military training, in operational situations, AR generally offers the most value. In particular, "situational awareness" tools can give personnel crucial real-time information related to their surroundings and position, plus whether those in the vicinity are friends or foes. As we'll see in this section, AR can even give military personnel improved night vision and potentially pave the way for superhuman vision in future.

Pilot displays

The use of heads-up and helmet displays in aircraft is nothing new. But the latest generation of helmet-mounted displays are incredibly advanced. One example comes from the F-35 Gen III helmet mounted display system by Collins Aerospace, which gives pilots "intuitive access" to flight, tactical and sensor information for revolutionary situational awareness. Serving as the pilot's primary display system, the system offers impressive virtual capabilities that enable the pilot to "see through" the bottom of the fuselage and look directly at targets for target verification. There's even a night vision mode, which eliminates the need for separate night vision goggles.

"Glass" tanks?

One of the upsides of being in an armored tank is you're relatively safe. But one of the downsides is it's not easy to get a thorough picture of what's going on outside – not without sticking your head out the top, which isn't ideal in a dangerous situation.

BAE Systems, maker of the formidable CV90 combat vehicle, plans to address this with AR capabilities and make a metaphoric "vehicle of glass"[vii] – a vehicle that maintains all the usual defensive capabilities, but is equipped with sensors and AR imaging systems that allow the troops traveling inside the vehicle to view what's going on outside, as if the tank were entirely transparent. In other words, using AR headsets, troops will be able to get a 360-degree view of the battlefield, identify threats and engage foes, all from within the vehicle.

Identifying and neutralizing bombs

The US Army has trialed AR goggles for combat dogs, which allow trainers to give remote commands to military dogs as they can scout for explosives and other hazards. Currently, the handler guides the dog using hand signals or laser pointers, but the handler needs to be close by for this to work. Using goggles, dogs can be guided to specific spots using on-screen visual cues, with the handler watching via video feed and giving commands from a safe distance. The goggles can be adjusted to fit each individual dog, and as military dogs apparently already wear protective goggles in bad conditions, it's only the AR commands that they'll need to get used to.

When it comes to disarming bombs, teams routinely use robots like SRI International's Taurus Dexterous Robot. Traditionally, robots like this would be controlled using a 3D monitor and remote controls. But today, the Taurus robot's arms and graspers can be controlled using an Oculus Rift headset, indicating an interesting new direction for bomb disposal technology.

Medical care in the field

What do you do if a fellow solider is injured but there aren't any medics in the vicinity? Or what if you're a medic tending to a wounded soldier, but the nearest surgical tent is miles away? With AR, military personnel and medics can respond to medical emergencies more efficiently, and even be guided by medics remotely.

That's the goal of US Army researchers, who are developing an AR system to help examine and treat injured personnel in remote environments. The AR surgical visualization software would allow medics to "see" a patient's internal anatomy, much like a CT scan in a hospital setting. (Or Superman's x-ray vision.)

And as for remote guidance, a study led by Purdue University has shown medics can successfully perform surgeries in battlefield-like simulations by receiving guidance from surgeons via an AR headset.[viii] The remote surgeon sees a video view of the patient and can mark up the images with drawings of how to conduct the surgery – and these annotated instructions are visible to the first responder within their field of view. The technology, called System for Telemonitoring with Augmented Reality (STAR) has been shown to help first responders with little or no experience successfully complete common procedures. Circle back to Chapter 7 for more healthcare examples, including the use of VR therapy for injured personnel and those with PTSD.

Improved situational awareness for troops

For troops on the ground, AR displays and glasses can deliver better situational awareness – meaning they can locate their position precisely, locate others around them and accurately identify whether others are a friend or foe. Such displays can also show other information, such as the distance to target, bringing to mind images of superhuman, Terminator-like soldiers.

In 2020, the US Army announced it was investing in 40,000 pairs of mixed-reality goggles (which is enough to outfit 10 percent of soldiers).[ix] Derived from Microsoft's HoloLens technology, the IVAS (Integrated Visual Augmentation System) goggles display critical information and are designed to help troops identify enemy forces and make decisions more quickly. Plus, the goggles are equipped with thermal and low-light sensors to help the user see in the dark. What's so important about systems like this is they are far more intuitive and easier to use than, say, handheld systems, which can distract the user. With goggles, the user never has to take their eye off the battlefield.

Eventually, IVAS goggles could incorporate facial recognition capabilities, or link with weapons to give soldiers the ability to fire without even seeing the enemy – meaning the soldier can "see" through the sight of their weapon using their goggles, so they can, for example, hold their weapon around the corner of a building and accurately fire, without having to stick their head around the corner. Troops will therefore be able to safely fire while remaining under cover.

Cyborg soldiers of the future?

Building on this idea of augmented soldiers, a report from the US Department of Defense (DoD) reveals that, within decades, the US military could create "machine humans," with superhuman vision and brains wired directly up to computers. The report, titled *Cyborg Soldier 2050: Human/Machine Fusion and the Implications for the Future of the DoD*, says that ear, eye, brain and even muscular enhancement is "technically feasible by 2050 or earlier."[x] Such technology could even allow troops to control unmanned vehicles with their thoughts and share data between personnel via two-way brain-to-brain interactions.

There would be a lot of ethical concerns and questions to iron out before the DoD's vision could become reality, but it's certainly an interesting sign of where things might be going in future. Read more about the future of XR in Chapter 13.

Lessons We Can Learn from Law Enforcement and Military

Generally speaking, the use cases in this chapter perhaps don't translate seamlessly to other industries. But here is what I've learned from this chapter:

- One key lesson we can glean from these examples is the incredible versatility of AR – particularly when it comes to the consumption of real-time data. In any industry, the ability to access, understand and act upon data is a vital part of business success.

Could AR help your teams use data more effectively, whether that is maintenance or assembly data, training data for new hires, product data, or any other kind of business information?

- This chapter also highlights some big ethical questions around the use of AR in certain situations. For example, should police on the streets be able to access facial recognition data via AR glasses, when the vast majority of people around them are not doing anything wrong? Sure, CCTV cameras are increasingly present on our streets but these generally aren't used to identify individuals' identities in real time. Shouldn't we have a right to individual privacy when we're just going about our everyday business? This is something Western countries in particular will have to grapple with.

- There are also huge questions around the potential to merge man and machine to create cyborg-like solders. Would such soldiers be more machine than human? How does the issue of user consent fit in with the military? Would ocular AR implants that give soldiers superhuman vision be reversible? Are we heading for a world in which certain sections of society are augmented while others aren't (a new class system for the twenty-first century)? Attempting to answer these questions is clearly beyond the scope of this book – I could write a whole book just on this issue alone – but I hope it serves as a warning to always keep ethics in mind when developing new XR applications.

Key Takeaways

In this chapter, we've learned:

- While VR can play a very valuable role in police and military training (circle back to Chapter 6 for more on training), it's AR that is proving more useful in tactical and operational situations.

- In law enforcement, key uses of AR include identifying suspects in real time (by combining AR glasses with facial recognition technology) and enhancing criminal investigations by overlaying critical information and instructions for first responders at a crime scene (thereby safeguarding evidence).

- Military uses of AR include heads-up displays for pilots, identification and disposal of explosives (including AR goggles for military dogs), improved visualization inside armored vehicles, better medical care in the field (by allowing remote medics and surgeons to provide visual instructions), and improved situational awareness in combat situations.

- The US Department of Defense has even outlined a vision of "cyborg" soldiers taking to the battlefield by 2050 – complete with superhuman vision and the ability to share data between teams using their thoughts.

That wraps up and concludes the real-life use cases. Now let's take a glance into the future and explore what may be coming our way in the years to come.

Endnotes

i. Police Augmented Reality Technology; FBI; https://www.fbi.gov/file-repository/stats-services-publications-police-augmented-reality-technology-pdf/view
ii. Military Augmented Reality Market to 2025 – Global Analysis and Forecasts by Components, Product Type & Functions; Research and Markets; https://www.researchandmarkets.com/reports/4471794/military-augmented-reality-market-to-2025#:~:text=The%20military%20augmented%20reality%20market%20is%20estimated%20to%20account%20for,US%24%20511.8%20Mn%20in%202017.
iii. Chinese AR start-up develops smart glasses to help police catch suspects; South China Morning Post; https://www.scmp.com/tech/start-ups/article/3008721/chinese-ar-start-develops-smart-glasses-help-police-catch-suspects

iv. US Services Testing VR-Based Flight Training Tools; Halldale Group; https://www.halldale.com/articles/15752-us-services-testing-vr-based-flight-training-tools

v. Martin Recruiters Want Cutting-Edge VR Flight Simulators to Attract Pilots; Military.com; https://www.military.com/daily-news/2020/05/25/marine-recruiters-want-cutting-edge-vr-flight-simulators-attract-pilots.html

vi. US Navy tests TRACER augmented reality combat training platform; Naval Technology; https://www.naval-technology.com/news/us-navy-tests-tracer-augmented-reality-combat-training-platform/

vii. AR warfare: How the military is using augmented reality; TechRadar; https://www.techradar.com/uk/news/death-becomes-ar-how-the-military-is-using-augmented-reality

viii. Augmented reality tool shown to help surgeons remotely guide first responders in battlefield-like scenarios; Purdue University; https://www.purdue.edu/newsroom/releases/2020/Q3/augmented-reality-tool-shown-to-help-surgeons-remotely-guide-first-responders-in-battlefield-like-scenarios.html

ix. "Mixed Reality" Goggles Will Give U.S. Army Soldiers Super Vision; Popular Mechanics; https://www.popularmechanics.com/military/a30898514/mixed-reality-goggles-army/

x. Cyborg warriors could be here by 2050, DoD study group says; Army Times; https://www.armytimes.com/news/your-army/2019/11/27/cyborg-warriors-could-be-here-by-2050-dod-study-group-says/

13
A LOOK INTO THE FUTURE

Back in Chapter 1, I briefly outlined a future in which the lines between the real world and the virtual world become increasingly blurred. In this final chapter, I delve a little deeper into that vision of the future, exploring where XR technologies are headed, and what this may mean for us as humans.

Rapid Advances in XR Technology Are Coming Our Way

XR interfaces are constantly evolving and, in the future, it's likely we'll experience XR in ways we can't yet imagine. But, for now, there are plenty of imminent tech advances to look forward to. We'll have faster, lighter, more affordable VR technology – including accessories that bring realistic sensations, like touch, to the VR experience. Advances in smartphone technology (such as better cameras) will mean we can enjoy slicker AR and VR experiences on our phones. And with 5G wireless networks, we'll be able to enjoy them wherever we are in the world. Here are some of the key advances on the horizon.

LiDAR: Bringing more realistic AR creations to our phones

The latest phones and tablets from Apple and other vendors are now equipped with LiDAR technology, which is intended to boost the devices' AR capabilities. But what is LiDAR? A LiDAR (Light Detection and Ranging) system is made up of a laser, which emits pulsed light, and a receiver, which measures the amount of time it takes for the light to bounce back – all of which can be used to create a 3D map of surroundings.

LiDAR can be used in various different functions (robot vacuum cleaners and self-driving cars use them to understand their surroundings), but it's LiDAR's AR-boosting abilities that are of most interest here. In a phone or tablet, LiDAR helps the device build a better picture of whatever the camera is pointed at, and this, in turn, helps apps add AR creations in a more realistic way. Crucially, because LiDAR creates a 3D map, it can provide a sense of depth to AR creations – instead of them looking like a flat, sticker-like graphic. It also allows for *occlusion*, which is where any real physical object located in front of the AR object should, obviously, block the view of it – for example, people's legs briefly blocking out a Pokémon GO character on the street as they pass by it. This is vital for making AR creations appear more rooted in the real world, and will avoid clunky AR experiences where objects appear to be in the wrong place. LiDAR may also bring virtual try-on apps for clothing into the mainstream, because it can map the body more accurately than a standard camera and therefore provide a realistic view of what an item would actually look like on the customer.

Bottom line, although lasers and sensors may not sound all that exciting, the incorporation of LiDAR systems into everyday phones and tablets could prove a major breakthrough for AR.

New advances in VR headsets

As well as becoming smaller, lighter and generally more comfortable to wear, VR headsets are just beginning to incorporate new built-in features to enhance the VR experience. Hand detection and eye tracking are two prominent examples.

The Oculus Quest 2 headset is the first to come with hand-tracking technology, but we can expect other mainstream providers to follow suit pretty soon. What's great about hand tracking – which accurately reflects the user's hand and finger movements – is it allows VR users to be more expressive in VR and connect with their game or VR experience on a deeper level, because they can control movements and other elements without the need for clunky controllers. Hand tracking could also prove to be a major stepping stone to social VR platforms and VR meetings becoming more mainstream, since it allows us to communicate with more natural gestures, just as we would in real life.

Eye tracking will no doubt prove to be another key VR milestone. In Chapter 3, I talked about how eye tracking works (and the potential privacy implications of tracking people's eye movements), so circle back there for a quick refresher. The HTC VIVE Pro Eye was the first mainstream headset to incorporate eye tracking technology (although, with a price tag of around $1,600, "mainstream" may be a bit of a stretch). It's reasonable to expect more affordable headsets, like the Quest 2, will adopt this technology in time. Why? Because, despite the privacy challenges and potential pitfalls, eye tracking helps to deliver a much smoother, more refined VR experience. For one thing, the system can focus the best resolution and image quality only on the parts of the image that the user is looking at (exactly how the human eye does). This taxes the system less, reduces lag and reduces the risk of nausea. And when combined with hand tracking technology, eye tracking could create an even more immersive, intuitive VR experience.

New VR accessories will hit the market

As well as slicker headsets, we can also expect a raft of external VR accessories and hardware to hit the market, all designed to be used alongside headsets to stimulate our senses. These will go beyond the sights, sounds and basic vibrations provided by current VR headsets and controllers, and make VR feel much more real.

One of my favorite examples is robotic boots. Yes, you read that correctly, robotic boots. If you think about it, VR experiences where you move around a virtual space can feel limiting – you can't physically walk around your real-life room to match your virtual wanderings because you'd end up bumping into walls and furniture! This is why some people experience nausea when they spend time in VR – their eyes tell them they're moving, but the brain knows the body is standing still. Startup Ekto VR plans to solve this through wearable robotic boots that provide the sensation of walking, to match your movement in the headset, even though you're actually standing still. Ekto One robotic boots look a bit like futuristic roller skates except, instead of wheels, they have rotating discs on the bottom, which move to match the direction of the wearer's movements. As you walk forward in VR, the boots pull your leg back, giving the sensation of physically walking. In future, accessories like this may be considered a normal part of the VR experience.

If you don't fancy strapping on robotic boots, another way to solve the motion problem is through omnidirectional treadmills – treadmills that keep you in one place while your feet slide around the treadmill. If this sounds a little too much like the movie *Ready Player One*, and nothing like reality, check out the Omni One home VR treadmill from VR startup Virtuix. The Omni One is a small, slippery, circular treadmill, complete with a body harness that holds your body in place while your feet slip across the platform – and the movement of your feet translates into the VR environment.

What about other sensations, particularly touch? We already have things like haptic gloves and suits that can provide haptic feedback, which simulates the feeling of touch through vibrations. (Read more about haptics back in Chapter 3.) In fact, full body suits are already available – the TESLASUIT being one example – but they aren't exactly achievable for everyday VR users. That said, they will become more affordable, mainstream and effective in time, providing yet another leap forward for VR. In short, haptics will make VR feel all the more immersive and realistic – for example, if you're playing a zombie game, you could feel zombie hands grabbing at your arms as you run away. Just imagine the impact this will have on action games . . .

Taking touch to a whole new level

Looking further ahead, beyond haptic suits (which can be clunky and uncomfortable to wear), developers are already working to create the next generation of wearable haptic solutions. These will bring an even more realistic sense of touch to VR experiences, even replicating human touch. It could even end up feeling better than human touch – or at least, create a different kind of human touch and interaction.

In one example, scientists at Northwestern University have been working on an experimental haptic patch that provides the sense of human touch across long distances. The lightweight, wearable, wireless patch can deliver a feeling like physical touch, corresponding to another user's movements. So, a parent at work could gently pat a VR interface, while a child at home wearing a patch on their back could feel the comforting patting. Crucially, the patch is only a couple of millimeters thick and flexes easily, so it can be comfortably worn directly on the skin. Technology like this could even integrate with social media and chat apps in future, so you can give your partner a loving hand squeeze or gentle stroke even when you're apart. Flirting over messaging apps could certainly get a lot more interesting!

Similarly, at the École Polytechnique Fédérale de Lausanne in Switzerland, researchers have created a soft, flexible artificial skin capable of delivering haptic feedback. Complete with integrated sensors and actuators, the membrane essentially inflates and deflates to provide the sensation of touch, and has so far been successfully tested on human fingers. Technology like this could also be used to create the feeling of touching different objects in VR.

In addition, there are projects under way to bring the sense of smell to VR experiences. One is the Feelreal mask prototype, which promises to incorporate the sense of smell, as well as other haptic sensations like heat, wind, water mist, and vibrations. Feelreal's "scent generator" component is comprised of replaceable cartridges that hold aroma capsules, capable of generating scents such as gunpowder and lavender. The makers intend to produce aromas that tie in with those used in the food industry – which could revolutionize travel- and hospitality-based VR experiences. In the future, technology like this could potentially be incorporated into off-the-shelf VR headsets.

Amazingly, smell could even be incorporated into AR. Scientists at Cranfield University in England have created a conceptual design for smell-based AR, which could allow AR experiences to enhance the user's sense of smell. Needless to say this is a long way off, if it even comes to fruition, but it's interesting to think that AR (not just VR) could become more stimulating and immersive. As I said right at the start of this book, we must remember that XR is a spectrum, and the distinction between the different XR technologies will no doubt become blurrier in future.

Integrating XR Technlogy into the Human Body?

VR skin patches are one thing. But how else might XR technologies integrate more seamlessly with the human body?

One way is through augmented reality contact lenses. While it's true that AR glasses will get better, cheaper and more comfortable, in the future they may also become obsolete as AR lenses take over. That's the (excuse the pun) vision of California-based startup Mojo Vision, which revealed in 2020 that it was developing AR contact lenses with micro-LED displays that place information inside the wearer's eyes. At the time of writing, Mojo's lenses are at prototype stage, and clinical trials will take years to complete, but the US FDA has already granted them breakthrough device status (designed to create a faster path to market for breakthrough medical devices), which is encouraging.

Imagine the uses for such AR lenses. For now, Mojo says that its first priority is to help people struggling with poor vision (by providing better contrast or the ability to zoom in on objects). But the intention is the lenses will eventually be made available for everyday consumers, and could be used to project things like health tracking stats and other useful data. Indeed, when demonstrating the prototype to journalists, the lenses displayed pre-loaded information such as text messages and the weather report, indicating that AR lenses could help us consume content in new ways. It could also help us enhance our sight in low light conditions (even if our vision is otherwise trouble-free), or even serve as a teleprompter for speaking events. Beyond that, the technology could eventually lead to us individually augmenting our own view of the world around us, as I mentioned at the start of this book. Don't fancy the idea of your vision being constantly disturbed? Thanks to image recognition technology, Mojo says the lenses will be able to understand what activities the user is engaged in, and therefore not disturb them at inopportune times.

This merging of the human body and technology brings to mind the super-human cyborg soldiers of Chapter 12. Could everyday citizens really be walking around like the Terminator, zooming in on objects far away and receiving information straight to our eyeballs? Assuming Mojo's lenses are just the beginning, it's certainly a possibility.

Another vision-based technology in its infancy is bionic eyes – functional implants for the eyes. This merging of man and machine isn't so far-fetched when you consider the use of bionic limbs for amputees. Why not bionic eyes as well?

Bionic eyes are intended as a future solution to poor vision or blindness (much like cochlear implants can restore a person's hearing), but, like the Mojo AR lenses, bionic eyes could potentially be marketed to everyday consumers in future. Various bionic eyes are already in development, including the Argus II Retinal Prosthesis System, which is designed to restore sight for patients with age-related macular degeneration. The system combines a retinal implant with glasses fitted with a camera – the camera takes pictures of whatever the wearer is looking at, then those images are converted into signals that are wirelessly transmitted to the retinal implant. From there, incoming information is transmitted to the optic nerve, and the brain then processes the information, allowing the wearer to "see" the image. Early tests with the Argus system show people can perceive shapes, but not colors, but the technology is in its early days.

If the thought of eye implants or AR lenses freaks you out, hold on to your hat. Because the next frontier in human–machine interactions is hooking our brains directly up to computers or VR experiences.

This is the aim of "neural VR." The basic idea of neural VR is users will be able to manipulate objects and control movements in the virtual world through thoughts alone. In other words, the VR systems of the future could harness brain waves to create a whole new level of immersion for users.

Boston-based startup Neurable is already working in this field, creating sensors that can decipher brain activity, understand the user's intention and then translate that into virtual reality. This is possible because every motion that we perform in the real world – walking, picking up a mug of coffee, even typing this sentence – produces brain waves. By recording and interpreting these brain waves, VR systems can produce the corresponding action in the virtual world. This may also have huge implications for neuro-rehabilitation; the combination of VR and brain–computer interfaces could potentially restore movement to physically impaired patients or guide patients through tailored rehabilitation exercises.

This notion of brain–computer interfaces – where external devices can be controlled using the brain – is the holy grail in the tech world. Elon Musk's Neuralink project is developing implants that will allow two-way communication between the human brain and a smartphone app, and the company hopes to start and expand human trials soon. Neuralink's goal is to help people with severe loss of brain function, but, if successful, it could well see wider adoption. Facebook is also developing its own brain–computer interfaces that can decode speech directly from the brain.

With names like Facebook and Elon Musk throwing their weight behind this technology, it's probably only a matter of time before it becomes reality. If integrated with XR technologies, it would revolutionize XR experiences and possibly change the entire nature of XR. What I mean by that is, in the future, brain–computer interfaces could be used to project images, sounds and the feeling of touch directly into the brain – so we can trick our brains into seeing and feeling what isn't there. Which would ultimately make external XR technology, well, redundant.

This is years, potentially decades, away of course. And there are obviously huge ethical challenges to overcome before we ever get to this point. But as a snapshot of what may lie ahead, it's certainly compelling. If I'm honest, I find the idea of brain–computer interfaces both creepy and exciting. The possibilities of, for example, restoring sight to blind people, or giving us all amazing night vision, are pretty cool. But there are potentially big downsides to hooking ourselves up to interfaces like this. Do we really want Facebook knowing our every thought? I don't. Then there's an even bigger question to address: will this merging of man and machine challenge our very perception of what it means to be human?

So Where Do We Go from Here?

It would be easy to paint all this with a dystopian flair – a slippery slope that starts with AR contact lenses and ends with humans permanently wired up to the Matrix! But I feel hugely positive about the future of XR. Sure, there will be ethical challenges to overcome and difficult questions to answer, but that goes with the territory for any new technology. And as with any new technology, we can't roll back innovations (almost everything mentioned in this chapter is already in development). How we use it is up to us. We must choose to use XR in the most ethical way possible, for the maximum benefit of everyone. We must avoid technology for technology's sake, in other words.

And, make no mistake, the potential benefits are significant – far outweighing the challenges, in my opinion. Certainly for businesses, XR offers huge scope to drive business success, whether that means engaging more deeply with customers, creating immersive training solutions, streamlining business processes such as manufacturing and maintenance, or generally offering customers innovative solutions to their problems. I hope the examples in this book have whet your appetite and tempted you to experiment with these fascinating new technologies – or at least learn more about them. True, some of the use cases within these pages verge on the gimmicky, but most are truly inspiring, particularly when it comes to improving education and healthcare. As I said at the start of this book, XR is about turning *information* into *experiences*, and this can make so many aspects of our lives richer and more fulfilling. That's why I believe XR has the potential to benefit society as a whole, for our generation and the generations to come.

Share Your Thoughts

I'd love to hear what you think about XR and its applications. Much as I enjoyed writing this book, I'm very keen to establish a dialogue beyond the confines of these pages. So feel free to ask questions, share any of your own XR success stories, or get in touch if you need help leveraging future technologies.

You can connect with me on the following platforms:

LinkedIn: Bernard Marr
Twitter: @bernardmarr
YouTube: Bernard Marr
Instagram: @bernardmarr
Facebook: facebook.com/BernardWMarr

Or head to my website at www.bernardmarr.com for more content and to join my weekly newsletter, in which I share the very latest information.

ACKNOWLEDGMENTS

I feel extremely lucky to work in a field that is so innovative and fast moving, and I feel privileged that I am able to work with companies and government organizations across all sectors and industries on new and better ways to use the latest technology to deliver real value – this work allows me to learn every day, and a book like this wouldn't have been possible without it.

I would like to acknowledge the many people who have helped me get to where I am today. All the great individuals in the companies I have worked with who put their trust in me to help them and in return gave me so much new knowledge and experience. I must also thank everyone who has shared their thinking with me, either in person, blog posts, books, or any other formats. Thank you for generously sharing all the material I absorb every day! I am also lucky enough to personally know many of the key thinkers and thought leaders in the field, and I hope you all know how much I value your inputs and our exchanges.

I would like to thank my editorial and publishing team for all your help and support. Taking any book from idea to publication is a team effort, and I really appreciate your input and help – thank you Annie Knight, Kelly Labrum, and Debbie Schindlar.

My biggest acknowledgment goes to my wife, Claire, and our three children, Sophia, James, and Oliver, for giving me the inspiration and space to do what I love: learning and sharing ideas that will make our world a better place.

ABOUT THE AUTHOR

Bernard Marr is a world-renowned futurist, influencer and thought leader in the field of business and technology. He is the author of 19 best-selling books, writes a regular column for Forbes and advises and coaches many of the world's best-known organizations. He has 2 million social media followers and was ranked as one of the top five business influencers in the world by LinkedIn.

Bernard helps organizations and their management teams prepare for the 4th Industrial Revolution that is fueled by transformative technologies like extended reality, artificial intelligence, big data, blockchains, cloud computing and the Internet of Things. He has worked with or advised many of the world's best-known organizations, including Amazon, Microsoft, Google, Dell, IBM, Walmart, Shell, Cisco, HSBC, Toyota, Nokia, Vodafone, T-Mobile, the NHS, Walgreens Boots Alliance, the Home Office, the Ministry of Defence, NATO, the United Nations, among many others.

Connect with Bernard on LinkedIn, Twitter (@bernardmarr), Facebook, Instagram, and YouTube to take part in an ongoing conversation, subscribe to Bernard's podcast, and head to www.bernardmarr.com for more information and hundreds of free articles, white papers and e-books.

If you would like to talk to Bernard about any advisory work, speaking engagements, or influencer services, please contact him via email at hello@bernardmarr.com

Other books in his Wiley series include:

Tech Trends in Practice: The 25 Technologies That Are Driving the
4th Industrial Revolution

Artificial Intelligence in Practice: How 50 Successful Companies
Used AI and Machine Learning to Solve Problems

Big Data in Practice: How 45 Successful Companies Used Big Data
Analytics to Deliver Extraordinary Results

INDEX

enhancing product design in
manufacturing 198–201
entertainment and sport 139–64
EON Sports 154
European Organization
for Nuclear Research
(CERN) 102
EvolVR (virtual meditation) 117
Expeditions (Google) 99
extended reality (XR)
definitions 5, 11–13
EYE-SYNC tool 120–1
eye-tracking 38–9, 120–1

F-35 Gen III helmets 224
Facebook 49, 239–40
Facebook Horizon 50–1
Facebook's Spark 79–80
FaceCake 80–1
fake news 37
fear of treatment, children
and 122–3
fear-of-heights simulators 125
Federal Bureau of Investigation
(FBI) 217–18
Feelreal headsets 28, 236
field trips 98–9
Firstborn (design and innovation
agency) 64
fitness coaching 118–19
FLAIM Systems 104–5
food industry 193–4
see also agriculture
Foot Locker 64
football 26, 107, 152–4
see also soccer

Ford 199–200
Fortnite 85–6
Froggipedia 103
FundamentalVR (surgical
simulators) 114

galleries see museums
gameChange project 125–6
gaming 25–6, 85–6, 143–6
Gap Inc. (apparel and
accessories) 71
Gatwick airport 190
GE (General Electric) 206–7
Gearmunk (gear reviews) 57
GearVR 16
General Electric (GE) 206–7
gigs/concerts 4, 155–8
glasses/goggles 13, 26–9, 120–1,
204–10, 225–7
global pandemics see COVID-19
Go Virtually 127
Google ARCore platform 29, 48
Google Arts & Culture 102
Google Cardboard 15, 16, 26, 28
Google Earth 183–4
Google Expeditions 99
Google Glass 13, 26–7,
204, 205–7
Google Street View 25,
26, 189–90
Google Translate 48–9
Grand Seiko (watchmaking
company) 79–80
Guided Meditation VR
116–17
guided visits see tours